POTS PANS PONIES & PINES

Cast Iron Cooking on the Trail

Marty Strannigan Coe

Copyright 2018 Marty Strannigan Coe

ISBN# 978-1-941052-33-4 Trade Paper

Library of Congress ControlNumber:
2018959884

Cover Design: Antelope Design

PronghornPress.org

Table of Contents

Dedicated to my grandmother, who was my anchor, an excellent cook, and who guided me through early life challenges. And to my mother, who let me wash horse halters in the washing machine and shared her love for the mountains with me. And to all my past, present and future Mountain Friends!

Cast Iron Cooking on the Trail

Me riding Hobbs.

Introduction

My name is Marty Strannigan Coe and I was born in 1950 in Ft. Collins, Colorado.

My dad was a basketball coach at Colorado State University in Ft. Collins. My mother was a stay at home mom with three girls and a new baby boy which was a houseful. I had two older sisters; one was seven years older and the other was five years older than me. I was four when my younger brother was born. Fortunately my grandparents lived close by and I was allowed to walk down the street to spend afternoons with them. This enabled my mother to have more time with the new baby and it allowed me to enjoy the smells and wonderful dishes that my grandmother would make every day.

At a very young age, I loved being in my grandmother's kitchen and started to help with small tasks. Mattie—"Grandma" to me—had very cool metal pans. They were triangular, similar in weight to cast iron and could easily be stacked in the cupboard. One interchangeable handle worked for all sizes. She must have brought them from Scotland as I haven't seen anything like them since.

Cast Iron Cooking on the Trail

Bacon drippings were stored in a large can on top of the gas stove. Each day the burners were lit with a match before meal preparation began. A pull-out pastry board was stored under the counter and above the flour and sugar bins. It was a different time: no fast food, no convenience food, just meals that were prepared from ingredients stored on the kitchen shelves.

I began making pies at an early age and loved eating the leftover cooked strips of piecrust that we sprinkled with cinnamon and sugar. We also made lots of shortbread and scones that we enjoyed with our afternoon tea.

My grandparents and dad were born in Scotland. They came to the United States when my dad was four. They didn't have a lot of material things, but I remember they continued to have afternoon tea, a Scottish tradition. Grandma would pour the tea into beautiful china cups and then add milk and sugar. If Grandpa was there to join us, she would put out slices of meat pie. I still have her cookie jar and teapot in my kitchen today.

Grandma was always sharing cooking tips; how to roll out the pie dough evenly and how to feel the heat of the oven before putting the food in to cook. I heard warnings about the consequences of not washing and rinsing dishes with hot water, how not to leave food out when it should be in the refrigerator and many other helpful suggestions. To this day, I still remember her advice when I cook in the mountains. It was a happy time for me and an early connection to the joy of cooking.

And then we moved. My dad was hired as the head basketball coach at Iowa State in Ames, Iowa. It was a good job, but my mother missed the Colorado mountains. We all missed the relatives we had left behind, so we decided to spend part of our summers at my uncle's cabin in Estes Park, Colorado. Of course there were still visits to my grandparents and Grandma's kitchen in Ft. Collins.

My uncle's cabin was pretty rustic. It had an outhouse and an outdoor shower. I remember taking baths in a huge washtub that Mom filled with hot water from the stove. In many ways it was similar to camping, with the exception we had a roof over our heads.

The main attraction in Estes Park for me was the pony ride concession at the edge of town. Of course, It didn't take long before all of us had outgrown the ponies and we were looking at the horse stable at the end of main street. We felt pretty special when we started renting horses at Stanley Hotel's livery stable. I was always doing odd jobs around the cabin so I could pay to rent a horse for another hour. On any given day, I fell in love with a different horse. At that time, we could take the horses on trails without a wrangler. That would never happen in today's world because of all the liability issues.

When I was nine my dad was offered the head basketball coach position in Laramie, Wyoming. He had played at the University of Wyoming, was raised in Rock Springs, so he was going back to his old "stomping grounds." We would also be only an hour away from relatives in Colorado.

Of course we met many friends and families that Dad had known when he played there. A close friend of his and an UW alumnus had several ranches north of Laramie. His daughter and I were the same age and I started riding the ranch horses with her. The more I rode, the more I wanted to ride.

I knew I wanted and *HAD TO HAVE* my own horse. After many hours of begging, we bought Torch: $1/2$ Thoroughbred, $1/2$ Appaloosa and "green broke." Obviously we weren't horse people, but he was mine.

Coincidentally, I had two girlfriends who lived in the same neighborhood on Spring Creek in Laramie. We spent the next six years, riding, getting bucked off and planning camping trips to Happy Jack. We rode our horses from Laramie to the summit, which was a pretty good trip for thirteen and fourteen year old girls.

Torch was instrumental in my move to Cody, Wyoming. While I was in college I took a summer job at Pahaska Tepee, a tourist resort at the East Entrance to Yellowstone Park. I knew they had horses there, so I thought I could bring mine. It never occurred to me that I should leave him behind in Laramie.

I packed my clothes, borrowed a horse trailer from our Laramie rancher friend, a truck from the University of Wyoming stadium crew, grabbed two friends and we were off.

When we arrived, I was told only horses that were part of the horse concession could stay on the Pahaska lease. Luckily I had a friend in Cody who would let me pay for pasture. But Pahaska Tepee was fifty miles west of town and I didn't have a car. As fate would have it, the manager owned a car and we shared the same day off.

Hank gave me a ride into Cody so I could ride Torch. He played golf on his day off and then we both went back to Pahaska to work. We became good friends. We enjoyed working, camping and hiking around Pahaska that summer. At the end of the summer we went back to the University of Wyoming to finish college. We had met in June and we were married the following March!

It would be another eight years before my love of cooking and the thrill of riding would come together to make a happy trail.

Why Dutch ovens?

Why Dutch ovens and cast iron skillets? They are extremely efficient when you're using coals. The coals heat the pans evenly and retain the heat for a long time.

Why *not* Dutch ovens and cast iron skillets? They are heavy to pack; sometimes there are fire bans and minimal wood supplies. There are now pans made from other materials that are advertised as being lighter and similar to cast iron. I have not tried them. No matter what sort of pans appeal to you, I would encourage you to at least try Dutch oven cooking at some point. If you are cooking for a group, it is a good way to get people involved. Part of the preparation for cooking is getting the fuel for coals.

For whatever reason, people in town hate to chop wood; but in the wilderness, they love it. I think it has something to do with the two-man saw they use. The guys are pretty competitive and it is fun to watch how fast and how efficient they can be working at each end of that saw.

I had a woman on one of the trips who was going through a pretty nasty divorce. She chopped wood with such a vengeance I was a little afraid to let her go home.

Cast Iron Cooking on the Trail

Guests love to build a fire, shovel coals over and under the pans and test to make sure food is cooking properly. Mountain magic kicks in! Jobs, news updates, traffic, and other worldly concerns are left at the trailhead. It is truly amazing to watch!

My favorite day is always one in the mountains or even just outside in nature. I love to hike, walk, weed, mow the yard, irrigate… well you get the idea. In rating mountain trips, I add a star if I am riding my favorite horse, I add a star if I have my favorite dog by my side and I add three stars if I'm cooking. Consequently, without much effort I get paid for a five star vacation.

Cooking is all about giving and receiving love. I love quotes and the following come to mind:

Cooking done with care is an act of love.—Craige Claiborne
The guests are met, the feast is set.—Samuel Taylor Coleridge
One cannot think well, love well, or sleep well, if one has not dined well.—Virginia Woolf.

For me, cooking in the mountains has always been a natural high. The campfire becomes the center of our mountain home. People gather around to get warm, share stories and of course eat the meal that has become a team effort. There are no newspapers and no cell service, but there are friendships to be made, physical challenges to be conquered and an appreciation of nature's gifts.

There has never been a bad meal because, in a way, it's a small part of the entire experience. Guests are always so appreciative and amazed by what can be cooked in a fire pit.

So much of the mountain experience is living with less. In today's gear mania, there are better tents and gear in general, but I still prefer to throw my "woods" bag with a cowboy canvas cover where I have a clear view of the evening sky. A bath consists of a dip in an icy cold stream, usually washing one limb at a time. A complete change of clothes in case you get wet, a few extra shirts, several pairs of socks and a warm hat is your wardrobe for a week. I am always sad when my last favorite pair of Levis suddenly has more holes than fabric.

On the trail you will experience drifting off to sleep counting falling stars, identifying the Big Dipper because that's the easy one, listening to a creek just steps away from your tent and enjoying leisurely dining while sitting on a wood stump.

It is always bittersweet on our last day when we ride to a spot on the trail where we can see traffic and hear civilization. If we have another trip booked before the end of the season, we feel pretty lucky.

I hope people will continue to use Dutch ovens and cook over the coals. It is so rewarding in the many ways I've tried to explain. Very simply, I know mountain dining feeds the body, but more importantly it feeds the soul and refreshes the spirit.

And remember: you don't need a pack string, wall tents and mountain meadows to try out the recipes found in this book. Open your kitchen window, get out your pans and mixing bowls and start cooking. An easy shortcut is to look at a traditional recipe and use the suggested time for cooking and oven temperature. Don't forget:cast iron will hold its heat longer than glass and aluminum cookware.

As an example, if I am cooking a roast, pork loin or chicken breasts, I pull the pan out of the conventional oven, tent it with aluminum foil and let it sit for an additional ten minutes. Soups, chili dishes and stews are fine to get out of the oven and place on the serving counter. A covered Dutch oven will keep the food warm until your family and friends are ready to "dish up." Desserts should cool without a lid.

Cast Iron Cooking on the Trail

In the Beginning

I was a stay-at-home mom with three children under the age of five. In pursuit of "me" time I would ask a friend to go riding with me on a Saturday or Sunday. We would ride a different trail on the North Fork of the Shoshone River each weekend. Then my South Fork friends invited me to ride with them on the other side of the ridge. I got very comfortable riding mountain trails, crossing rivers and streams and seeing an occasional bear or moose.

One day in the grocery store, I ran into a friend of mine who played football at the University of Wyoming. Joe DeSarro, who was an outfitter and managed Valley Ranch, took many pack trips up the South Fork. We talked about the mountains, beautiful trails around Cody and riding. The next thing I knew I was riding into Boulder Basin to help set up tents for his hunting camp.

From the very beginning I knew I didn't want to be the only woman in hunting camp. So when I was asked to help out, I would grab my sister or a girlfriend who enjoyed riding into the mountains. I always owned two good horses and had my own horse trailer to get us to the trailhead. The outfitters were happy with that arrangement

Cast Iron Cooking on the Trail

because they had one more person to help and they didn't have to provide the horses.

It wasn't long before I became good friends with Joe's wife, Sally. We shopped, cooked and packed together for many years, on many trips.

On one particular trip, Joe assigned the cooking duties to my sister and me. That was the first of many meals prepared over the fire. I always remember that trip because of the frozen (rock hard) hamburger. We had packed the pannier with enough food for one night and two days. The hamburger was at the bottom and was still frozen when we rode into camp that night. The first lesson learned: always pull the dinner meat out and put it on top of the pack if it's frozen.

We put it in a bucket of hot water and it did thaw out. There are many times in the mountains when you just have to "make do" with what you have to work with.

I've cooked in five different hunting camps owned by four different outfitters. I was always a substitute cook and several times just went in to help the outfitter's wife. Good outfitters have great wives who help with menus and equipment lists for the cook, sometimes shop for groceries, and sometimes go in to cook.

Joe DeSarro's camp was a ten mile ride into Boulder Basin. They called it "Colder Boulder." Enough said!

Joe Tilden's camp was twenty-eight miles into the Thorofare. We rode up a steep trail with many switchbacks, over Deer Creek Pass and into a great hunting area called Hidden Basin. Many times the pass still had snow drifts at the top in July and August. On one trip we had to dig thru a huge drift to get over the pass and into camp.

Digging thru a drift in the middle of a blizzard to get farther into a wilderness camp seemed like a crazy plan. But the hunters did well, and by the time we rode out, the trail was clear and it was a beautiful autumn day.

Tim Hockhalter's camp was located eighteen miles up Timber Creek in Crandall. Although the scenery was awesome, the trail had many sharp turns and long stretches of sliding shale.

Lee Livingston's camps were located on the North Fork.

POTS, PANS, PONIES & PINES

Camp Monaco was a ten mile ride, Elks Fork about eight miles in, and Jones Creek was six miles from the trailhead. They were all easy, good trails.

I feel so very fortunate that I worked with great outfitters who knew how to pack and work with horses and mules.

Riding different trails is one of the challenges of being a camp cook. Weather is another one. In the fall the hunters like a good storm because it gets the elk moving down to feed, but it can make for long days and cold nights. Also, if it gets too warm and they kill an animal, the wranglers have to make a meat run back to civilization to prevent spoilage.

Cooking and packing is actually easier for hunting camp. Typically the hunters come for 7 days. The menus are the same for each week during the season. Cook for 7 days, have 2 days off, then start another 7 days with a new group of hunters. The guests are there to hunt, so that is their passion, mission and entertainment. Cooking and getting the meals into the dining tent is the cook's job description.

The cook has a camp wrangler who cuts the wood, hauls water and keeps the cook stove going. It takes a meal or two to figure out how much wood to use and how to regulate the heat on the stove, but after a couple of days, it's pretty easy.

Hunters like to come to the cook tent to wind down, get warm and talk about their families, but most of the time they are talking to their guides.

Most outfitters buy groceries for the entire hunting season. Since we had set menus, we would meet at the barn and pull the food that we needed that week. The staples: dishes, pans, coffee/spices/cake mixes, etc. were left in the cook's tent for the next group. We basically just had to restock meat and perishables. The wranglers packed the food.

Hunters were always easy to cook for, but I know it was the only time I have been asked to scrambled elk brains and boil elk tongue!

Cast Iron Cooking on the Trail

Me, Lee and Sally.

Lee Livingston

Lee Livingston was the owner of the fourth hunting camp where I cooked. I spent most of that time in Camp Monaco, which was originally Buffalo Bill's Hunting Camp. It got its name because Buffalo Bill once hosted the Prince of Monaco there on a combined hunting and fishing trip.

I first met Lee when Sally and I were cooking on a return trip from Jackson Hole. We were riding back with guests who took a trip with the DeSarros every year. We had ridden for a few days and our last camp out was in Simpson Meadows after we dropped over Marston Pass. We were tired and in a hurry to unsaddle so we could get dinner started.

I saw one of the older wranglers untying my horse, Papigo. I immediately stopped him and told him they always used Papigo (which was actually one of Sally and Joe's horses) for a picket horse. They're called a "picket" horse because the animal is tied to a long

Cast Iron Cooking on the Trail

rope hooked to a metal stake hammered into the ground. The picket horse is the leader of the herd. The other horses are released but will stay in the same general area of the meadow if the picket horse remains tied and the feed is good. But if the picket horse is released, he will take the rest of the herd down the trail. Sometimes they use two picket horses and hope that some of the horses will stay if one is left behind.

In this case the wrangler didn't like me telling him what to do. In fact, he said, "Why don't you do what you know best, and that would be cooking."

He released my horse and Papigo took all the other horses down the trail with him.

Since Lee Livingston was only eighteen and the youngest wrangler on the trip, he was immediately sent down to bring all the horses back. I would add that he was sent down the trail without help. Wranglers had a pecking order and you had to work your way up that ladder. This was a "wrangler test!"

We were just hoping he would catch up with them and bring them back before dark.

When we woke up the next morning, Lee and the horses still were not back. We had all our gear and guests to take out with half the stock we needed. We double saddled the few horses we had left and started walking down the trail to Purvis Meadow. We were still a long way from the trailhead. A couple of miles down the trail, here came Lee with eighteen horses all lined out. What a sight to behold! Our Hero! The horses had gone all the way out to the trailhead—about eighteen miles—and Lee had ridden most of the night and brought them back at daylight.

That older wrangler didn't release my horse again.

Lee worked for Joe DeSarro's outfit for several years. Sally and I got to know him really well and we became friends. About the time Lee bought his own camp, Sally and Joe moved to Montana. I continued to cook for Lee. I would ask Sally to come down to help and sometimes I would ask another friend.

POTS, PANS, PONIES & PINES

When I was fifty-three, I was divorced. I wanted to keep busy and I needed to be in the mountains. There were several times I couldn't find a friend to go with me, so I started to cook by myself. To make me feel more comfortable, Lee would come into the cook tent and make an attempt to talk "girl talk," but he really wasn't very good at it. So, I got used to flying or riding solo, as they say.

Initially, I would cook for eight to ten people on four to five day trips. Lee actually had changed the menus to have less Dutch oven cooking and more skillet meals. For example, we rarely had prime rib on the menu, but we always had spaghetti and meatballs.

An exception was when Lee asked me to cook for twenty-two people on a Cody to Jackson trip. I had cooked for that group before and knew they were very particular about their food, especially dinner. I was nervous about the preparation as well as the cooking, but I couldn't find anyone to help so I agreed to go solo. I took my horse, Warren, my dog, Chancey, and hoped for the best.

The organizers of that trip had specially cut veal chops prepared specifically for their guests. Everyone had been talking in anticipation for that meal. I put foil on a pannier lid, seasoned the meat and took it over to the grate. At the time I was thinking to myself that the meat probably cost more than my mortgage payment, but everything went okay. It was a good trip and good for my spirit.

The next fall Lee asked me to cook for three weeks in hunting camp. Once again I wondered if I could do everything that needed to be done by myself, but I was still trying to figure what the rest of my life was going to be, so I agreed. I have to say at that time in my life I was saying "yes" to any new adventure. I even said yes to surf camp on the North shore in Hawaii. I failed miserably at surfing, but I didn't say no to a new adventure.

Lee Livingston's camp was the easiest "ride in" of all the camps. We trucked the mules and horses to the trailhead at Pahaska Tepee, Buffalo Bill's original hunting lodge. Lee had two camps which were called Jones and Monaco that were located on the North Fork of

Cast Iron Cooking on the Trail

the Shoshone River. I always cooked at Monaco, which was an easy ten mile ride.

Eric, the camp wrangler, was from West Virginia. He liked to whitewater raft and decided to try another adventure. He was a youth counselor in his "real life." We had so much fun, and as with many mountain friends we still stay in touch.

Wyoming wranglers are pretty tough. For example, at one breakfast I served quiche. Lee said in a loud voice, "Cowboys don't eat quiche."

And I responded, "Real cowboys don't need women to cook for them." We were always teasing each other.

I think it was the second hunt when Eric shared that he'd had a massage in Cody on his day off. I mentioned it in the dining tent around the other wranglers. Everyone stopped talking and just stared at Eric. He really got teased from then on for getting a massage on his day off. Most of the other wranglers spent their off time in town at Cassies, the local cowboy bar.

Since our camp wasn't very far into the wilderness, we were able to work for six days, ride out that afternoon, have a day in town to shower, restock, etc. and then ride back in on the eighth day with a new group of hunters.

Sometimes the ride in and out was miserable. I remember a trip when it was snowing and raining and the thunder was echoing down the valley. It was a deep wet kind of cold. I had worn so many layers of clothing that I had trouble climbing on the horse, but I was still cold. On those bad weather days, I learned to just lower my head and let my horse take me down the trail.

A Day in Hunting Camp

When the cook rides in with the first group of hunters, she brings enough food for the first hunt. She also brings in the staples (coffee, salt, etc.) and most of the ingredients for baking, enough to last the entire season. Wranglers take the game meat out at the end of the hunt and bring back items the cook needs.

The cooking tent is usually set up adjacent to the dining tent, although some camps combine the two. It is set up as close to a stream or river as possible. In some camps I slept in the cook tent, although I don't think they allow that now. I always thought maybe it wasn't a good idea to be frying chicken in my bedroom, especially when there were bear tracks around the stream.

In hunting camp, they have one wrangler whose job is to chop wood, bring in water, light fires in the tents in the morning and generally help the cook. I remember in Joe Tilden's camp, I would wake up to a handsome wrangler lighting my fire so my tent would be warm before I got out of my sleeping bag. He would even bring me a hot cup of coffee. It was wonderful! I thought I had died and gone to heaven. Everyone worked hard to make the cook happy. I think they just didn't want to have to cook themselves!

Cast Iron Cooking on the Trail

In camp mornings start about 4:30 a.m. The hardest part of a new day is the short time in the morning when the tent is still cold. I would be mixing up pancake batter and looking at the ice crystals on the top of the tent. Luckily, it wouldn't take long for the cook tent to heat up and soon I'd be peeling off layers.

Breakfast goes pretty fast with the hunters wanting to be out of camp by 5:30 a.m. Some days I would go back to bed for an hour or two after they left. Most of the hunters grabbed snacks to fill their saddlebags, never knowing when they might be back. Most mornings I would fix a pot of soup in case some of them came back early, have lunch makings readily available and maybe fix a dessert for the evening meal.

There is more time in hunting camp to write in journals or read books and magazines. Preparation for dinner starts around 3:00 in the afternoon. All baking is done on a Sims Stove, a small stove that can be broken down and packed in by mules. It was just like cooking at home but with better scenery.

If a hunter kills an elk, the celebration is on. In one camp they asked me to cook the tongue, which I did. I basically boiled it forever, pealed it, diced it, covered it with garlic and sautéed it. It was very "chewy." They would also want the elk brains scrambled in their eggs the next morning.

Each camp has their own traditions that include passing around the Crown Royal bottle or another favorite "poison." As the week goes on, more hunters are able to kill their elk and then they will stay in camp. Wranglers and hunters who stay in camp spent time chopping wood, repairing hitching rails and tack, and playing cards.

Most of the men I met talked a lot about their kids and wives. I was asked more than once if I was nervous being in camp alone with lots of men. But really, if you didn't have antlers on your head, they didn't pay much attention to you…that is until they got hungry or needed a fresh pot of coffee.

Every group is different and every trip is different. The following list is a basic list of equipment and kitchen dishes, etc. that outfitters pack into the mountains.

Equipment list:

Implements:

3-4 mixing bowls
1 large coffee pot, 1 small coffee pot
1 measuring cup
1 set measuring spoons
1 can opener
Pot holders
Towels, dish towels
2 metal water buckets for heating water
2 six gallon water containers
2 metal dishwashing pans

Utensils:

2 paring knives
2 butcher knives
2 rubber spatulas
2 metal spatulas, at least one with long handle
2 long handle forks for cooking
tongs
2 wooden spoons
whisk
small cutting board

Cast Iron Cooking on the Trail

Pots & Pans:

1 large 14 qt. Dutch oven, 1 small Dutch oven
1 large cast iron frying pan
1 Teflon frying pan
2 Jackson Grills, grates
  (1 is enough for 6 to 8)
2 grill guards
2 saucepans
2 qt. Pitchers

Serving Items:

2 large serving spoons
1 corkscrew
dinner plates
coffee cups
soup bowls
platic juice glasses
knife/fork/spoon place settings

Miscellaneous:

water filter
first Aid Kit
water bottles
  (for those who forgot theirs!)
lanterns
fly wipe
flashlight
foam pad
  (usually cut up to protect cinch sores)

POTS, PANS, PONIES & PINES

Grocery Store Items:

paper plates
Handy Wipes
Chore Girl Scrubber
paper towels
napkins
lunch bags
aluminum foil, large heavy duty, small-regular
liquid hand soap
dish soap
trash bags: one for each day and a couple extra
Ziploc bags: gallon, quart and sandwich
small SOS pads
Handy Wrap
toilet paper: at least one for each day and a few extra
waterproof matches
fire starter

Cast Iron Cooking on the Trail

Seven Day Hunting Camp Menu

Pack lunches for the ride in:
*Note: Bring cooked turkey from home
to serve on the first night*

Day 1

Dinner

Turkey/Dressing
Mashed potatoes
Pea Salad
Cranberries
Rolls
Pumpkin pie/coffee

Cast Iron Cooking on the Trail

Day 2

Breakfast

French Toast
Scrambled eggs/bacon
Juice/Coffee

Lunch

Sandwiches/Soup

Dinner

Spaghetti with meatballs
Garlic Bread
Caesar Salad
Lemon Bars/ Coffee

Day 3

Breakfast

Skillet Omelette
Skones
Juice/Coffee

Lunch

Sandwiches/Soup

Dinner

Ham
Garlic Grits
Baked beans
Corn Bread
Pineapple Upside Down Cake/Coffee

POTS, PANS, PONIES & PINES

Day 4

Breakfast

Pancakes/sausage
Fried eggs
Orange juice/coffee

Lunch

Soup/Sandwiches

Dinner

Stew with biscuits
Coleslaw
German Chocolate Cake/Coffee

Day 5

Breakfast

French toast/ham
Eggs
Juice/coffee

Lunch

Soup/Sandwiches

Dinner

Stuffed pork loin
Applesauce
Tossed salad
Cheese cake/coffee

Cast Iron Cooking on the Trail

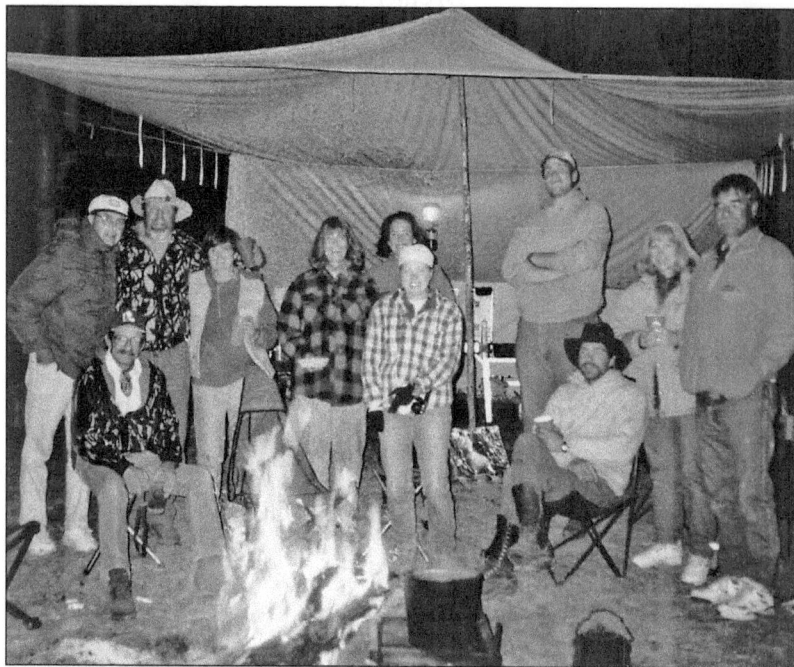

POTS, PANS, PONIES & PINES

Day 6

Breakfast

Pancakes/Bacon
Scrambled eggs
Juice/Coffee

Lunch

Soup/Sandwiches

Dinner

Fried chicken
Corn
Mashed Potatoes
Apple Crisp/Coffee

Day 7

Breakfast
Bagels
Eggs
Sausage patties
Juice/ Coffee

Pack lunches for ride out

Cast Iron Cooking on the Trail

Sally relaxing on a layover day.

Sally

I met Joe's wife Sally DeSarro at a children's play group sponsored by the church we attended. We had children the same age and we both shared a love for horses, cooking, adventure, and for the mountains.

We became very good friends and have maintained that friendship through lots of life changes. Joe and Sally started to organize summer pack trips for the ranch owners and their friends. There were several trips to Boulder Basin, which was an easy ride in and out.

At some point, they decided it would be fun to ride to Jackson Hole. We rode from Valley Ranch, up the South Fork trail, over the Buffalo Plateau and down to Turpin Meadows. We stayed two nights and one day in Jackson. We spent the only day there buying groceries, collecting other needed supplies and repacking. We also had to pick up the group that we would be taking back to Cody. Some years my oldest son and Sally and Joe's children came on the return trips.

The first year we did the Cody to Jackson and back to Cody trip, we booked rooms in a five star hotel in Teton Village. We did not know there was a black tie event for the Teton Symphony held in the

Cast Iron Cooking on the Trail

village that night. There are not words to describe the hotel guests' shock when they saw our dirty, smelly group check in with guns and a few saddles! A cowboy does not leave his saddle at a trailhead. Not only are they expensive, but it takes some time to find one that fits you like a glove. And gun laws were not so strict thirty-five plus years ago, so pretty much everyone carried a gun. We must have made too much noise that night, as we were asked to relocate the next day.

Most of the Jackson trips came off without too many mishaps. However one year, Sally and I decided to have additional "town fun" and pack at the trailhead. Most years we packed the panniers in our hotel rooms. We would grab old newspapers from the sidewalk stands to pack the metal cans tight so they wouldn't rattle when riding down the trail. We had become confident in our packing abilities, so we decided to take the groceries to the trailhead and pack there.

The trailhead was twenty miles from town. We unpacked the groceries, made our piles for each day's menu and stood back. It didn't look right. We were missing two full days of groceries. We didn't have a choice. We had to drive back to the grocery store.

We found two full carts of food. The clerk knew we'd be back. As luck would have it, we were also missing two horses. By the time everything was found, we were on the same page again with the wranglers. When you work in the mountains everyone gets used to doing a lot of problem solving. And we agree to live by the code, "nothing is a big deal unless you make it one!"

The length of a summer trip and the trail taken depend on the guests. Some people like to ride over passes, some like to have a basecamp and just take day rides and some like destination trips. We've taken guests to the Beartooth Mountains, Dubois, Jackson and Yellowstone Park. We have done the Cody to Jackson trip and returned on different trails. We have never had a bad trip, but we have had some interesting experiences.

Sally and I cooked well together. She would start doing a salad while I got the meat ready. I would get the appetizer table ready and she would be chilling the dessert. We knew the menus really well and knew how to work efficiently to get everything done.

POTS, PANS, PONIES & PINES

One year we were hired by a dude ranch between Cody and Thermopolis. They raised cattle. The guests and crew took the cows up to the high pasture in the spring and then had a different group of guests that brought the cows down in the fall. We didn't have to buy groceries for the trip, so we loaded up our horses and met everyone at the ranch.

It was a very unusual situation. They didn't have split wood. They had built several new cabins and had leftover plank wood, which doesn't burn well and really doesn't make coals. They didn't have enough food for the guests, so they had to make additional trips to town. The first night was really challenging.

Also, we didn't know anyone. We spent that first night sleeping in the back of my truck with cows mooing on one side and wranglers and guests partying on the other. We really didn't sleep that well that night and wondered what we had gotten ourselves into.

The next day we started up the mountain. The ranch had two chuck wagons pulled by teams of horses to carry the food up to the high mountain meadow. The guests kept the herd moving upward while Sally and I brought up the rear to gather any cows straggling behind. Everything was going fine until the team ran away with one of the wagons.

Whatever had spooked the horses caused them to gallop up the mountain, dangerously rocking the chuckwagon back and forth on its axels. The cowboy with the reins had to just hope the horses would run out of energy and slow down before there was a major wreck. We were lucky; it was a short chuck wagon with only two horses. We caught up to the wagon after the cowboy finally got the horses shut down.

We opened the back of the wagon and OMG what a sight! Flour, dishes, vegetables, broken eggs, spices, coffee grounds and coffeepots scattered every which way. But we got to camp, started cleaning up the mess and pretty soon we had dinner finished, dishes done and a tired bunch of guests in bed.

We met some interesting characters on that trip. Phillip was a farrier from Boston. A farrier is just another name for a horse shoer. He did a great job keeping all the stock shod and healthy. Phillip loved to sing and loved to dance. He would start singing when the sun

Cast Iron Cooking on the Trail

was coming up and continue until the fire was put out at night. Most of the crew and guests were from Boston. The ranch owners were from Boston and that is where they marketed their Wyoming ranch adventures. And it was an adventure! Lots of dancing, singing and story telling around the campfire! They must have liked our cooking because they hired us back in the fall.

I look back on those early trips and I think how lucky we were. Almost every meal was an experiment in working the fire, getting the temperature right and choosing menus that worked for different size groups, appetites and food preferences. I think we survived because we both enjoyed the work, the company and we didn't dwell on the difficulties.

Specific menus did rise to the top of our lists. We particularly liked Chicken Marsala. It was a spicy, creamy chicken dish that we served over rice. We started calling it Chicken Mar Sal.....for Marty and Sally, of course!

Cast Iron Cooking on the Trail

Summer Pack Trips

Cast Iron Cooking on the Trail

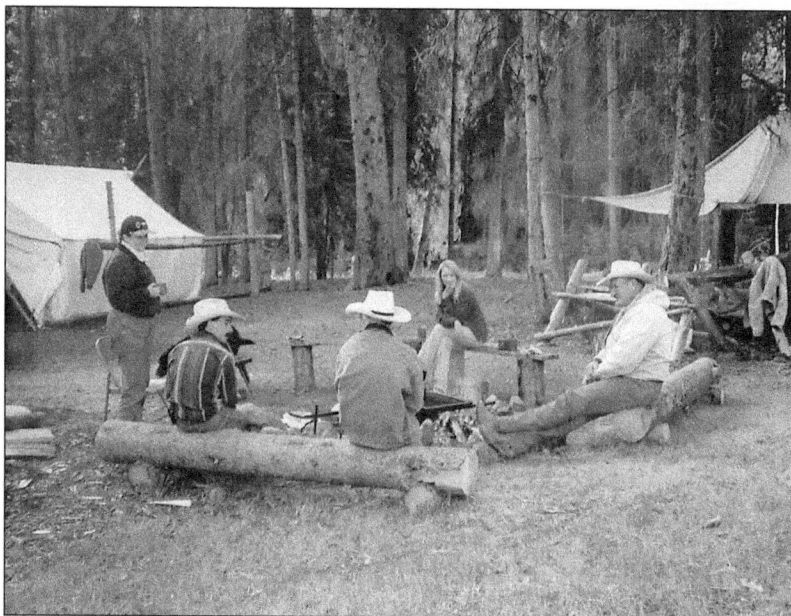

POTS, PANS, PONIES & PINES

Summer camp trips, especially the moving trips are much more labor intensive. I would plan the menus, do the grocery shopping and pack the panniers the day before we left.

If Sally was coming with me, I would have help and of course that made it more fun. We needed to be at the trailhead usually by 5:30 a.m. in hopes of leaving by 8:00 a.m.

We would also bring sack lunches for the first day. If we were going into the Thorofare, we would stop for lunch on a slope before going over Deer Creek Pass. It was a beautiful spot, next to a creek. And I would usually find my favorite wildflower there: forget-me-nots. The wildflowers are beautiful in the backcountry. There were many trips where I hated to throw down my sleeping bag because I would be crushing bouquets of flowers.

Summer horseback trips do not have specific camp locations. Outfitters get a certain number for day use and then they can apply for different camps in different areas. They can also exchange day use days in certain areas with other outfitters.

Hunting camps are permanent. When a hunter draws a license in a particular area, then he chooses an outfitter to take him into the outfitter's camp in that area.

Guests that go in on summer trips call an outfitter and tell them what activities they want to experience. For example, if they are avid fishermen, we take them where the fishing is good. Or if they

Cast Iron Cooking on the Trail

want to ride different trails and see a lot of country, we plan a moving trip, which would go from camp to camp. There are a small group of outfitters that are approved annually to take trips into national parks.

On summer trips everyone pitches in to set up camp. The guests usually put up their own tents and carry their gear to their tents. Most guests volunteer to help with horses, wash dishes and offer to help in all areas. They are there for the adventure and don't want to be pampered.

The wranglers and the cook usually don't have a tent as they prefer to look at the stars and don't want the hassle of putting up another tent. The cook helps unpack the horses while the wranglers help put up the cook tent. Sometimes the crew sleeps in the cook tent or the cook fly if the weather gets bad.

The wranglers take care of the horses while the cook starts dinner. The fire is lit first as it takes lots of coals to cook dinner in a Dutch oven.

After the meal is prepared and buried in the coals, it's time for a glass of wine or favorite beverage. It is always fun to get to know the guests. The first night is a little different as people are tired and still nervous. By the end of the week, we have become very good friends and usually stay in touch the rest of our lives.

Of course there are many guests that are excellent riders, have owned their own horses and have had many adventures. They usually want to help the crew and of course their help is always appreciated.

Sleep comes early in the mountains, usually around 9:00 or 9:30.

Sally and I would get up at 5:30 in the morning to start breakfast and pack lunches if it was a moving day. On the days we didn't move, we would sleep in until 6:30.

Breakfast was a good time to warm up around the fire and drink lots of camp coffee. Even in the summer, high mountain camps make for chilly sleeping. Layover camp days were spent fishing, reading, journaling or taking short rides from camp.

One of my favorite times in camp was bath day. We would go down to the river and get in the ice cold water. It was usually a "wash a limb at a time" type of bath. We would never just dive into a deep hole.

After bathing, we would spend another hour drying our hair, filing our nails and soaking up the sun. In a very short time, we would be back cooking in the camp smoke, but the bath was certainly therapeutic. Our favorite camps usually had our favorite bathing spots.

Warren, my favorite mountain horse and Hobbs's best friend.

Cast Iron Cooking on the Trail

Me leading a pack horse and a pack mule

Summer Horse Pack Trip
Menus for Seven Day Moving Trip

Friday: Departure Day

Pack lunches
Dinner: Peanuts, cheese and crackers, steak,
baked potatoes, tossed salad, rolls,
pound cake with strawberries and whipped cream.

Saturday: Layover Day

Breakfast: Pancakes, eggs, bacon
Lunch: Taco Salad
Dinner: Snow peas with ranch dressing,
grilled pork tenderloin, rice pilaf, steamed carrots with parsley,
cornbread, apple crisp with caramel sauce

Sunday: Travel Day

Breakfast: Eggs, sausage links, French toast
Lunch: Pack Lunches
Dinner: Cheese and crackers, smoked oysters, hamburgers, fried
potatoes, coleslaw, baked beans, cherry chocolate cake

Monday: Layover day

Breakfast: Skillet omelet, toast, blueberries, cantaloupe
Lunch: Chili with hot dogs
Dinner: Guacamole dip with salsa and chips, pretzels, chicken fajitas,
Spanish rice, refried beans, peach cobbler

Cast Iron Cooking on the Trail

Tuesday: Travel Day

Breakfast: Pancakes, sausage patties, eggs, oranges
Lunch: Pack lunches
Dinner: Curry dip with carrots, cheese and crackers,
spaghetti and meatballs, Caesar salad, garlic bread, cheesecake

Wednesday: Travel Day

Breakfast: French toast, bacon, eggs
Lunch: Pack Lunches
Dinner: Broccoli with ranch dressing peanuts,
fried chicken, mashed potatoes with gravy,
corn, rolls, lemon pie with raspberry topping

Thursday:Travel Day

Breakfast:English muffins, ham, eggs
Lunch: Pack lunches
Arrive at trailhead/end of trip

The Recipes

Cast Iron Cooking on the Trail

Breakfasts

Cast Iron Cooking on the Trail

"Bear-proofing" the panniers.

POTS, PANS, PONIES & PINES

HINTS: There are many good mixes (pancake, muffins, coffee cakes, biscuits, etc.) that are great timesavers. Some cooks prefer to cook from scratch so that's why I've included my favorite "scratch recipes." I use a combination of both.

If you have room, frozen orange juice is worth packing in. It also helps to keep the chill in your cooler.

To warm up the syrup, place the bottle in a bucket of hot water. I usually pack in one glass container and refill from plastic bottles if needed. That way you don't have to worry about a plastic container melting.

On summer trips, I use a "Jackson Grill," which is a square griddle that swings on a stationary metal stake. The stake is pounded into a corner of the pit, usually next to the grate. I've been told that back in the day, cooks used the top of a washing machine.

The temperature can be controlled by raising or lowering the grill closer to or farther away from the coals. Or it can be rotated to the side, which moves it entirely away from the fire. It is a wonderful addition to anyone's camp equipment.

A Jackson Grill can also be used as a lid for a large skillet. You also want to use it to put coals on top, which allows you to bake food in your large skillet as you would a Dutch oven. Always wrap the Jackson Grill in foil if you are using it as a lid. This will keep it

Cast Iron Cooking on the Trail

clean and ready to use for cooking pancakes, French toast, and other "grilled" things.

In hunting camps, I use a large flat griddle. By moving the griddle around the top of the woodstove, you can find the hotter and cooler spots. The hottest spot for cooking is usually over the firebox and also toward the back of the stove.

A large Dutch oven works well to keep things warm. I line the bottom with foil and then put a couple of paper towels on top of the foil. I place breakfast meat on the bottom and put pancakes, muffins or toast on top. I put the Dutch oven to the side of the grate. You want a little warmth from the fire to keep the food warm while you finish cooking.

If you're cooking on summer trips, you probably won't have a stove. When you get to the camp, a wrangler usually helps dig out the fire pit. I like a fire pit that is shaped like a "T" and is about 3 feet long and 2 feet wide.

I put the cooking grate at the top of the "T." This is where I keep the coffee pot and fry the breakfast meat. To check to temperature of the coals, I hold my hand a few inches from the heat source. Very hot, (400 degrees) would mean you couldn't hold your hand there very long. Low heat (250) would mean you could keep your hand there much longer.

You never want to cook over a hot, open flame fire. Start your wood at the bottom of the "T" and wait for the coals. Use your shovel to get the coals from under the burning wood to your cooking spot and add more wood to the fire. So you are burning the wood on one side of your cooking pit and cooking on the other side. Wood that has been split in smaller pieces makes better coals for cooking.

Don't be afraid to experiment! For example you might find a large rectangle pit works better for you and your group. If it is a larger group, you sometimes have a grate at each end of the fire pit with the "feeder fire" in the middle.

Be sure to take care of the camp. Always place the sod and dirt to the side of the fire pit so you can replace it when you are ready to move to the next camp. When you leave, make sure the fire is out and you replace the sod and dirt.

We always try to leave starter wood for the next group that will be coming down the trail.

Kris, Tim Hockhalter, and me.

Cast Iron Cooking on the Trail

Breakfast Fruit

Try to put a small amount of fresh fruit out for every breakfast.
Pick one of the following:
sliced cantaloupe
green and/or red grapes
orange slices
apple slices
plums
nectarines
(canned fruit is fine, if fresh is not an option)

Heart Mountain Camp Coffee

Fill coffee pot with cold water to desired level, but if filling to the top, leave 2 inches.

Spread coffee over the top of the water and then add an additional amount to have an approximately $1\frac{1}{2}$ inch coffee mound in the middle. It will look like a small Heart Mountain.

With an average size coffee pot, you will add one C of ground coffee. With different size pots, the Heart Mountain method seems to work.

Heat until water and coffee come to a full boil. Let it boil for several minutes. Remove from the fire. Pour 1 C cold water in coffee to settle the grounds.

You can reheat the coffee (and add to) during the day, but I like to make a fresh pot every morning for breakfast.

Cast Iron Cooking on the Trail

Breakfast Casserole

Brown 2 lbs. link sausage. Drain and cool.

In a 9 X 13 pan or skillet put 8 slices bread with the crusts removed. On top, grate $3/4$ lb. cheddar cheese, then add sausage.

Mix the following and pour on top:

6 beaten eggs, $2^1/4$ C milk, $3/4$ t mustard.

Cover with foil. If you want to make this ahead you can put it on the top of the cooler pannier the night before.
Wrap the Jackson Grill with foil so you can put coals on top. Put the skillet in the pit and bank the coals on the sides and top. Coals should be medium heat (similar to 350° in an oven). Cook for 30 minutes but check after 10 and rotate.

Muffin Tin Bacon and Eggs:

Cook bacon until it is half done.

In each muffin tin, put a small amount of butter.

Place $1/3$ slice bacon (partially cooked) in each tin and spoon 1 T salsa on top of bacon.

Break an egg on top of bacon.

Cook in medium to hot oven (375°) until half-cooked bacon is done. This recipe works well in hunting camp where you have a cook stove.

Skillet Omelets

Hint: Brown hash browns and sausage the day before for quick assembly in the morning.

In large skillet:
Brown 1 pkg. (2 lbs.) hash browns (or enough to cover bottom of skillet) until crisp.
Brown 1 lb. of sausage and put on top of hash browns. (Can also use chopped ham)

Add whatever other ingredients you want:
chopped onion
green and red peppers
grated cheese

Pour $1\frac{1}{2}$ dozen beaten eggs on top. Cover with foil.

Place foil covered Jackson Grill on top. Place in fire pit and bank coals on side of skillet and add coals on top of Jackson Grill.
Cook until eggs are firm. Rotate the skillet several times.
Approximately 35 minutes with medium coals.

Cast Iron Cooking on the Trail

Omelets in a Bag

eggs
chopped ham
chopped green and red peppers
chopped green onions
grated cheese
heavy duty quart size Zip-Lock bags

Crack 1-2 eggs into zip-lock bag and as many ingredients as you would like. Season with salt and pepper. Make sure the bag is sealed and then squeeze the bag to make sure the egg yolks are broken and the ingredients are well mixed.

Place bag in pot of boiling water. Be careful that plastic bag does not touch sides of the hot boiling pot. Several bags can be cooked at once.

Serve with fried scones. And don't forget to put out the salsa and Tabasco!

Try this recipe on a layover day when breakfast usually lasts a couple of hours.

Easy Breakfast Burritos:

Cook bulk breakfast sausage until brown; drain.
Scramble eggs.
Add sausage to cooked scrambled eggs.
Warm tortillas.
Put cooked eggs and sausage mixture in tortilla.
Serve with salsa.
Amounts will depend on the number of guests.

Mexican Breakfast Burritos

12 large eggs
1 t salt
$^1/4$ t pepper
2 T butter
$^1/2$ C chopped onion
$^1/2$ C drained canned diced tomatoes
$^1/2$ C chopped green bell pepper
12, 8-9 inch flour tortillas
$2^1/2$ C thick, chunky salsa

Beat eggs, salt and pepper in medium bowl.
Melt butter in heavy large skillet over medium heat.
Add onion, tomatoes and green pepper and sauté until tender (about 4 minutes). Add egg mixture and stir until softly set.
Remove from heat.

Wrap tortillas in foil and place on grate until warm. When warm, spoon egg mixture down the center of each tortilla.
Spoon 2 T salsa over eggs on each tortilla.
Fold tortilla sides over eggs, and then roll up tortilla to enclose contents completely.
Place burritos seam side down in skillet. Spoon remaining salsa over the top and sprinkle with cheese. Cover with foil.
Leave on the warm side of grate until cheese melts and filling is hot. About 15 minutes with medium coals.
These can be made ahead, frozen, and then heated through.

Some guests prefer to hike rather than ride a horse. If they want to get an early start, I make burritos. They put them in their day packs and hit the trail. You can also do this with the Breakfast Sandwich.

Mexican Creamed Chicken

Topping:

$1/2$ can diced green chilies
1 glove garlic
1 T cumin
1 T dried cilantro
1 T salt
1 T pepper
1 medium onion, chopped
$1/2$ lb. pork sausage
1 t cayenne pepper

Sauté all ingredients until sausage is brown. Set aside.

Cream sauce:

1 cube (8 T) butter
4 C milk
$1/2$ C flour

Melt butter in saucepan over low heat. Blend in flour. Add milk all at once. Cook quickly, stirring constantly until mixture thickens and bubbles.

Take 4 chicken breasts, cooked, skinned and cubed and add cubed chicken to cream sauce.

Brown the biscuits.
Spoon topping on each biscuit.
Pour cream sauce over both.

Breakfast Sandwich

Brown English muffins and ham on griddle.
Can also use bagels
Spread mustard on bread.
Add ham, slice of cheese and fried egg.

Mom's Biscuits and Gravy

Biscuits:

2 T baking powder to each cup of flour, pinch of salt.
1 heaping T shortening to each cup of flour.
Add milk enough to make moist.

Mix everything together but try not to handle too much.
Roll out on floured board. Cut out with top of glass or can.
Put in greased skillet. Cover with foil.
Top with foiled Jackson Grill.
Put in pit and bank with medium coals on sides and top.
Bake 10-12 minutes or until nicely browned.

Brown Sausage in skillet. Remove sausage and make gravy.

Gravy:
$1^1/_2$ C milk
3 T flour combined with 3 T sausage drippings.

Warm the milk, mix the flour and drippings in a separate cup, then add back to the milk.
Return sausage to pan and serve over biscuits.

Cast Iron Cooking on the Trail

Pancakes

Makes 6-8 average size

1 $1/4$ C flour	1 T sugar
2 t baking powder	$1/2$ t salt
1 beaten egg	1 T sugar

1 C milk (for thinner pancakes, add additional 2 T milk)

Mix together dry ingredients.

Combine egg, milk and salad oil. Add to dry ingredients, stirring just until moistened.

Bake on hot griddle.

Keep warm by lining large Dutch oven with foil and then putting pancakes in Dutch oven. Keep oven close to heat source.

Variations:

Blueberry pancakes: When undersides of pancakes are nicely browned, sprinkle about 2 t drained blueberries over each cake. Turn, and brown other side.

Buttermilk pancakes: Substitute buttermilk or sour cream for sweet milk. Add $1/2$ t soda and reduce baking powder to 2 t.

Sourdough pancakes: Substitute flat beer for milk.

NOTE: Whether cooking on a woodstove or over coals with a Jackson Grill, it is important to have the griddle the right temperature and level. It is worthwhile to adjust your wood before you start cooking. I always say I have to cook a couple for the dog (too gooey or too burned) before I really get rolling.

Pancake mixes that you only have to add water to, save a lot of time on busy days.

Yogurt Pancakes

Makes 6 to 8 average size

1 C flour
2 T sugar
1 t baking powder
$1/2$ t baking soda
$1/2$ t salt
2 eggs
$1^1/2$ C plain yogurt
2 T melted butter
$1/2$ C water

Mix all ingredients and cook on hot skillet

Cast Iron Cooking on the Trail

Grandma's Griddle Scones

2 C flour
$2^1/_2$ T sugar
$^1/_2$ cube butter
1 t baking soda
$^3/_4$ t cream of tarter

Mix all ingredients until butter is worked into flour.

Add enough buttermilk to hold the dough together and a good consistency to roll out.

Divide dough in half, roll out, then cut into four pie wedges.

Fry on a hot griddle.

Serve with honey or jam.

Can also be cut in half, (sliced thinner) and served with butter and syrup.

Variations

Cheddar-Thyme Scones:

After combining butter and dry ingredients, stir $1^1/_2$ C grated cheddar cheese and 1 T chopped fresh thyme into flour mixture before adding buttermilk.

Sprinkle the tops of scones with an additional $^1/_2$ C grated cheddar cheese.

Raisin Scones:

Add $^1/_4$ C sugar to dry ingredients. After combining butter and flour mixture, stir in 1 C raisins.

Theo's Make-Ahead Coffeecake

$1^1/_4$ C flour
$^1/_4$ C sugar
$^1/_4$ C shortening
$^2/_3$ C milk
1 T baking powder
1 egg

Topping:

4 T flour
3 T butter
6 T sugar
1 t cinnamon

Mix dry ingredients. Cut in shortening.
Add milk and egg.
Pour into 8 inch square pan or small skillet.
Sprinkle topping over batter.

Cover with plastic and place in top of cooler pannier overnight.
In the morning, remove plastic wrap, cover with foil and top with foiled Jackson Grill. Bank coals around sides and put on top of grill. Bake for about 25 minutes, but check and rotate the pan after 10.

Raisin Coffeecake

$^1/_2$ C butter or margarine
1 C sugar
2 eggs
1 t vanilla
1 C sour cream
2 C all-purpose flour
$1^1/_2$ t baking powder
1 t soda
$^1/_4$ t salt

Topping:

1 C broken walnuts
$^1/_2$ C sugar
1 t ground cinnamon
$1^1/_2$ C raisins

Cream together butter and sugar.
Add eggs and vanilla; beat well.
Blend in sour cream.
Mix together flour, baking powder, soda and salt. Stir into creamed mixture and mix well.

Spread half the batter in greased medium skillet.
Put half the topping on top of batter. Spoon on remaining batter. Sprinkle on remaining topping.
Cover with foil. Top with foiled Jackson Grill. Bank sides with coals and put additional coals on top.
Bake in medium coals for 40 minutes. Check and rotate pan after 20 minutes.

Breakfast Cookies

**This is a great wholesome recipe to make at home, freeze,
and then pack into the mountains.*

2 C old-fashioned oats
1$^{1}/4$ C whole wheat flour
1 C all-purpose flour
1 C Grape-Nuts cereal
$^{1}/2$ C wheat germ
$^{1}/2$ C bran
2 t baking soda
2 C (4 sticks) unsalted butter, room temperature
2 large eggs
1 C (packed) brown sugar
$^{1}/2$ C sugar
1 T vanilla extract
1 C almonds (about 5 oz.) toasted, coarsely chopped
1 C raisins (about 5 oz.)
1 C chopped pitted dates (about 8 oz.)

Preheat oven to 350.
Mix oats, both flours, Grape-Nuts, wheat germ, oat bran, and baking soda in large bowl to blend.
Beat butter in another large bowl until creamy.
Add eggs, both sugars, and vanilla. Beat until smooth.
Add cereal mixture, stir until blended.
Mix in almonds, raisins and 1 C chopped dates.

Grease cookie sheets. Using damp fingers, press cookies to $^{1}/2$ inch thick mounds. Bake cookies until brown on top, about 15 minutes

Cast Iron Cooking on the Trail

Lunches

Cast Iron Cooking on the Trail

POTS, PANS, PONIES & PINES

Moving Days

Cast Iron Cooking on the Trail

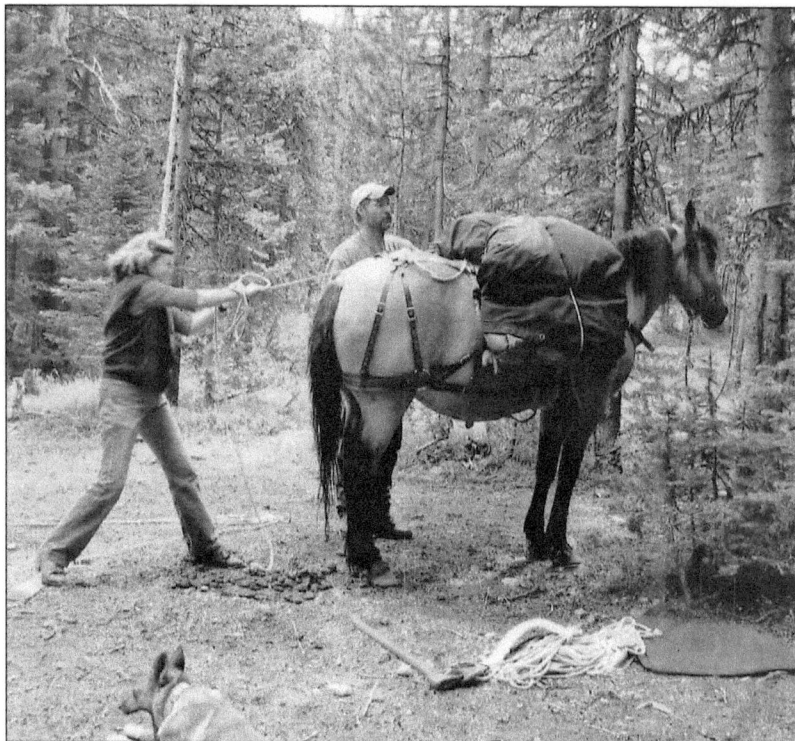

POTS, PANS, PONIES & PINES

The best thing to do for lunches on a moving day is to put everything that goes into a good sack lunch on a table and let the guests build their own. Guests know how much food they will eat and what they like to eat. They will be happier and you won't have a lot of wasted food.

The one exception to this is the first day on the trail. On that day I make lunches for everyone and include a bottle of water.

Items to include on the lunch preparation table:

Deli meat of the day (turkey, beef, pastrami or ham)

Cheese slices

Peanut Butter and Jelly

Bread: whole wheat and crusty white

Tortillas

Condiments

Fruit

Veggies: carrot and celery sticks, pepper strips

Candy bars or granola bars

Cookies

Hard Candy (Jolly Rogers or Lemon Drops)

Pringles
*I put a tube of Pringles in my saddle bag
and then share at our lunch stop*

Brown Lunch Bags

Sandwich Bags

Snack Bags

Juice Boxes

Cast Iron Cooking on the Trail

Wraps of all kinds are a nice change from the traditional sandwich. If you have leftover meat from the previous night's dinner, slice it thin and put it on the lunch preparation table. Romaine lettuce travels well and provides a nice crunch. Many guests prefer lettuce wraps. I have included a couple of recipes that guests enjoy.

Ham or Chicken Salad Wraps

1 C cooked chicken or ham, finely minced
$1/4$ C celery, finely minced
4 black olives, finely minced
4 T green pepper, finely minced
$1/2$ C mayonnaise or ranch dressing
4 flour tortillas-small

Mix all ingredients. Spread on tortilla. Tuck one end under and roll the tortilla over the filling.

Turkey Wraps with Curry, Chutney Mayonnaise and Peanuts

1 C mayonnaise
$1/4$ C mango chutney (store bought)
$1^1/2$ t curry powder
4 (9-10 inch) diameter spinach flour tortillas or plain flour tortillas
2 C diced cooked turkey meat
6 T coarsely chopped, lightly salted peanuts
$1^1/3$ C coleslaw mix
(preferably green and red cabbage with carrots) without dressing
1 C chopped fresh cilantro

Mix mayonnaise, chutney and curry powder in small bowl. Set aside.

Mix the turkey, peanuts and coleslaw in small bowl. Season to taste with salt and pepper.

Spread 2 t curry-mayonnaise in 2 inch wide strip down center of 1 tortilla.

Spoon turkey mixture on top of curry-mayonnaise. Sprinkle cilantro on top of turkey mixture.

Fold in one end of the tortilla over filling and continue to roll the tortilla into a tight wrap.

Cut filled wrap in half.

Cast Iron Cooking on the Trail

Waitng for a truck ride into Jackson

Layover Days

Cast Iron Cooking on the Trail

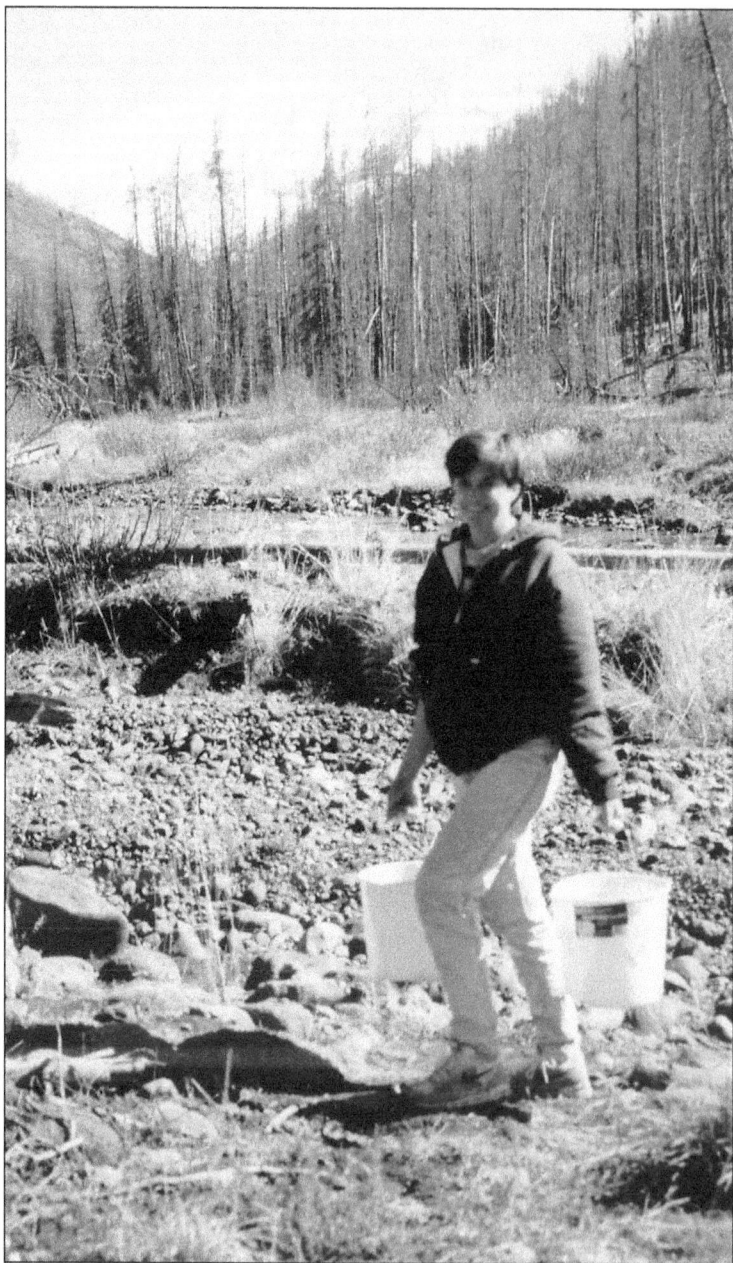

I love layover days. They are so relaxing. Since we are not moving, we get an extra hour of sleep, an extra cup of coffee over breakfast and usually a bath. Some guests like to take a short ride out of camp, some go fishing or hiking and some just hang out in camp.

I put out the lunch prep table for those who want to leave camp, but for those that stay in camp, I prepare another dish.

The following list of menu choices works well. You will find the recipes in other sections of this book.

For layover days and all dinners, make a large pitcher of ice tea and a large pitcher of lemonade or flavored fruit drink. Crystal Light is a brand we usually use. It makes a lot and doesn't take up much room in the packs. Place all drinks on the serving table. The serving table should also have a large bowl of fruit and should be available at each meal.

Chili – Skillet meals
Serve with Fritos corn chips and shredded cheddar cheese
Calico Beans – Skillet Meals
Serve with cheese quesadillas, cut in quarters
Taco Salad – Salads
Serve with extra tortilla chips and salsa
Cobb Salad – Salads
Serve with grilled cheese sandwiches
Macaroni and Cheese – Skillet meals
Option: slice in hotdogs
Onion Soup with Crusty Bread – Soups
Gazpacho* – Soups

Make this at home, freeze hard and it will still be cold when you get to camp. Serve with croutons, chopped cucumbers and green onions

Cast Iron Cooking on the Trail

Finger Steaks

1^1/$_2$ lb. round steak
Store bought marinade sauce
^{1}/$_2$ C ketchup
^{1}/$_3$ C vinegar
2 C flour
salt and pepper to taste
dash of Tabasco sauce

Marinate meat in marinade sauce for 1 to 2 hours. Drain.
Cut meat in finger strips ^{1}/$_4$ inch thick.
Mix ketchup, vinegar, salt and pepper, Tabasco sauce. Dip meat in mixture and roll in flour.
Sauté on both sides in skillet in hot oil. Cook 6 to 8 pieces at a time for 2 to 3 minutes or desired doneness.
Serve with hot barbecue sauce.

Crunchy Vegetable Burritos

^{1}/$_2$ C shredded carrots
^{1}/$_2$ C chopped broccoli
^{1}/$_2$ C chopped cauliflower
2 green onions, thinly sliced
4 oz. shredded cheddar cheese
^{1}/$_2$ C ranch dressing
^{1}/$_2$ t chili powder
4 (7 inch) flour tortillas
1 C lettuce, torn into small pieces

In a bowl, combine all but last two ingredients. Mix well.
Spoon ^{1}/$_2$ C veggie mixture onto tortillas. Top with lettuce

POTS, PANS, PONIES & PINES

Hot Dogs or Tasty Sausages

Talk to your butcher and buy the brand or type they recommend.
Grill on grate over hot coals.
Turn often.
Slice or put out buns with all types of mustard and pickles.

Cast Iron Cooking on the Trail

POTS, PANS, PONIES & PINES

Soups

Cast Iron Cooking on the Trail

There are very good "soup starters" that can be purchased and are very useful in hunting camp. They're easy to prepare and will not spoil if you don't use them.

I usually buy an assortment. I think my favorite is potato/leek, but I will buy bean, vegetable and chili. They are dried and come in foil packets that are easy to pack.

The weather can change in a hurry, so sometimes the hunters will be in camp most of the day. A hot lunch is appreciated.

I have also learned that buying a little extra food is a good habit. Many times we share a meal with weary backpackers or with a group that has forgotten a portion of their groceries at the trailhead.

Cast Iron Cooking on the Trail

Gazpacho

3 large tomatoes, peeled and chopped
1 green bell pepper, chopped
1 cucumber, peeled and chopped
$^1/_2$ C green onion chopped
4 C tomato juice
2 avocadoes, chopped
5 T red wine vinegar
1 small garlic clove, minced

Mix together.

Serve chilled with croutons, additional chopped cucumbers and green onions on the side.

I make this at home and then freeze it. I pack it in the cooler panniers. It is a pleasant surprise for guests in a summer camp.

Cast Iron Cooking on the Trail

Potato Soup

2 medium onions, thin slices
2 leeks, thin slices
5 medium sized potato, peeled and cubed
1 quart chicken broth
1 T salt
2 T butter

Sauté onions and leeks in butter.

Add potatoes and broth and bring mixture to a boil. Simmer for 35 to 40 minutes until potatoes are very tender.

Add 4 C milk (or 2 C milk and 2 C half and half)

If soup needs to be thickened, add instant potatoes, a T at a time, while mixture is hot.

Hunters love potato soup!

Variations:

Add canned salmon, sausage or cut up peppers.

On summer trips, I whisk the mixture very smooth, chill and then sprinkle chopped chives on top.

Quick Change Chowder

2 stalks celery, thinly sliced (1 cup)
3 T butter   2 T flour
1 carton (32 oz.) chicken or vegetable broth
2 C frozen diced potatoes with onions, partially thawed
$1^1/_2$ C milk (or more for desired consistency)
1 C half and half
4 slices bacon, crisp-cooked and crumbled
salt and ground black pepper to taste

In a saucepan, cook celery in hot butter over medium heat about 5 minutes or until tender.

Stir in flour. Stir in broth. Bring to a boil, stirring constantly.

Add potatoes, return to boiling. Reduce heat and simmer, uncovered 15 minutes or until tender.

Slightly mash potatoes. Stir in milk, half and half and bacon. Heat through.

Variations:

Make the above basic chowder recipe and add the following ingredients for different flavor options.

Seafood and sweet and spicy red pepper: Cook one red sweet pepper, seeded and chopped, and $1/_4$ t crushed red pepper flakes. Add 8 oz. poached, grilled or steamed salmon or tuna, cut into bite-size pieces

Artichoke, crab and basil: Add 12 oz. lump crab meat, drained and flaked, 1 (13 oz.) can artichoke hearts, drained and coarsely chopped and $1/_3$ C fresh basil or 2 t dried basil.

Chicken with peas and carrots: Add 1 C frozen peas and carrots, thawed, 8 oz. frozen chicken tenders (thawed, cooked and chopped) and 1 T fresh tarragon or 1 t dried tarragon.

Cast Iron Cooking on the Trail

Clam Chowder

3 cans (10.5 oz.) chopped clams
1 C chopped onion
1 C chopped celery
2 C cubed potatoes
salt and pepper to taste (lots of pepper)
3/4 C butter
$^1/_2$ C flour
1 pint half and half
1 pint milk
2 T Sugar

Pour clam juice on the diced, chopped and cubed vegetables and add water just to cover. You can substitute extra bottle (8 oz.) of clam juice for water for extra flavor.

Simmer for about 30 minutes or until vegetables are tender.

In separate pan, melt butter, stir in flour.

Add half and half and milk.

Add to vegetables.

Add clams last.

POTS, PANS, PONIES & PINES

Hank's Corn Chowder

1 large chopped onion
$^1/_2$ chopped green pepper
6 to 8 peeled cubed potatoes
4 to 5 cans (10.5 oz.) cream style corn

Sauté onion and pepper in small amount of butter.
Add potatoes and cover with 1 to 2 inches water.
Simmer until potatoes are tender.
Add corn.
Add enough half and half to the desired consistency.

Variation:

Add chunks of sausage or pieces of cooked bacon.

Broccoli and Cheese Soup

1 quart water (add chicken bouillon to taste)
1 quart half and half
1 C grated cheddar cheese
2 pkg. frozen chopped broccoli, thawed and drained
2 T butter
1 C chopped onions
1 C sliced mushrooms

Sauté onions and mushrooms in butter Set aside.
Boil water and add chicken bouillon (usually 1 bouillon cube
for each cup of water.) Add onions, mushrooms and broccoli.Simmer
for 15 minutes. Add cheese and half and half before serving.
Continue to stir until cheese is melted.

Vegetable Beef Soup

3 lbs. round steak cut in 1 inch pieces
1 bottle (1-pint 2 oz.) tomato juice
$1/3$ C chopped onion
4 t salt
2 T Worcestershire sauce
1 t chili powder
2 bay leaves
1 large can (1 lb. 13 oz.) whole tomatoes
1 C diced celery
1 can (10.5 oz.) whole kernel corn
1 C peeled sliced carrots
1 C peeled diced potatoes

In large Dutch oven or soup kettle combine meat, tomato juice, onion, seasonings and 6 C water.
Cover and simmer 2 hours.
Add vegetables, cover, and simmer for one hour.

Variation:

Add $1/2$ to $3/4$ C barley

Chicken Tortilla Soup

1 1/2 lb. boneless, skinless chicken breasts, cooked and cubed
1 C onion chopped
1 t McCormick's minced dry garlic
1 T oil.
3 C chicken broth
1 can (14.5 oz.) Mexican tomatoes
(diced fire roasted or spicy salsa)
1 T chili powder
1/4 t ground cumin
2 T cornstarch

Sauté onion and garlic in oil in medium Dutch oven.
Add chicken, broth, tomatoes and seasonings.
Bring to a boil and then simmer for 20 minutes.
Add cornstarch to thicken.

Serve with shredded Mexican Cheese Blend on top and sprinkle minced fresh cilantro on cheese. Sour cream and tortilla chips or warm corn tortillas can be served on the side.

Cast Iron Cooking on the Trail

Taco Soup

1 lb. ground sirloin beef
1 pkg. taco seasoning
1 pkg. ranch dressing mix
1 can (10.5 oz.) whole kernel corn
1 can (4 oz.) diced hot green chilies
1 can (15 oz.) chili beans
1 can (15 oz.) kidney beans
1 can (10 oz.) Rotel diced tomatoes and green chilies
1 can (14.5) diced stewed tomatoes
2 cans (24 oz.) water

Brown the beef in a soup kettle.
Add taco seasoning and ranch dressing mix while browning.
Add remaining ingredients to the soup kettle.
Bring to a slow boil. Add salt and pepper to taste.
Reduce heat and simmer for 30 minutes.
Serve with tortilla chips.

Joe's French Onion Soup

4 large onions, thinly sliced
2 T butter or margarine
3 cans (10.5 oz.) condensed beef broth
1 T Worcestershire sauce
salt and pepper to taste
French bread, sliced in 1 inch slices, buttered and toasted
grated Parmesan cheese
grated Swiss cheese

Cook onions in butter in medium Dutch Oven until lightly brown and translucent. About 15 minutes.

Add broth and Worcestershire. Bring to boil. Season with salt and pepper.

Place toast on top of onion broth. Sprinkle Swiss and Parmesan cheese on top of toast.

Place Dutch oven in warm fire pit. Place a few coals around the side and a few on top. Bake for 10 minutes.

Toast should be brown and cheese should be melted.

Hunting Camp Bean Soup

1 lb. dry navy beans
1 meaty ham bone
1 t salt
6 whole black peppercorns
1 bay leaf
1 medium onion, sliced
salt and pepper to taste

Wash beans. Add 2 quarts cold water; soak overnight. (Or, simmer 2 minutes; remove from heat; cover and let stand for one hour.)
Don't drain.
Add ham bone, salt, peppercorns and bay leaf.
Cover and simmer 3 to $3^1/2$ hours
Add the onion in the last half hour.
Remove bone, cut ham off bone and add to soup.
Mash beans slightly.

Minestrone Soup

2 cloves garlic finely chopped
1 medium onion, sliced thin
2 small zucchini, peeled and sliced in $1/2$ inch slices
2 medium carrots, peeled and sliced in $1/2$ inch slices
2 medium celery stalks, peeled and sliced in $1/2$ inch slices
1 large can (1 lb. 13 oz.) diced tomatoes
4 C chicken broth
4 C tomato juice
1 C water
salt, pepper, basil to taste
2 C uncooked spaghetti or shell pasta

Sauté garlic and onion until onion is tender.
Stir in remaining ingredients except pasta.
Heat to boiling, reduce heat, cover and simmer for 45 minutes.
Bring back to a boil and add pasta.
Cook until pasta is tender.

Cast Iron Cooking on the Trail

Hearty Minestrone

2 lbs. chuck roast
1 T salt
4 quarts water
1 celery, chopped
1 onion, diced
1 carrot, peeled and sliced
3 tomatoes, peeled and chopped
2 T minced fresh parsley or 2 t dried parsley
1 can (6 oz.) tomato paste
2 T salt
1 t dried oregano
$1/2$ t dried basil
1 can (15 oz.) dark red kidney beans
1 can (15 oz.) can garbanzo beans
1 can (15 oz.) can baked beans
1 pkg. (10 oz.) frozen chopped spinach, thawed
3 small zucchini, sliced
1 lb. Italian cooked sweet sausage, sliced
1 pkg. (8 oz.) uncooked, spaghetti or shell pasta

In large soup kettle, combine chuck roast, 1 T salt and water.

Cover and simmer for 3 hours or until beef is tender. Remove meat from kettle and set aside.

Remove fat from broth by letting kettle cool and then lift hardened fat off top and discard.

Break meat into 1-inch pieces and combine with broth in a very large stock pot.

Stir in celery, onion, carrot, tomatoes, parsley, tomato paste, salt, oregano, basil, kidney beans with liquid, garbanzo beans with liquid, baked beans, spinach, zucchini and sliced sausage. Simmer, covered, until vegetables and sausage are tender, about 1 hour.

At this point you can freeze the soup.

To serve, cook noodles according to package directions.
Rinse, drain and add to hot soup.
Sprinkle each serving with Parmesan cheese.

This recipe makes 8 quarts of soup. It is a good recipe to fix at home and then pack into hunting camp for the first night.

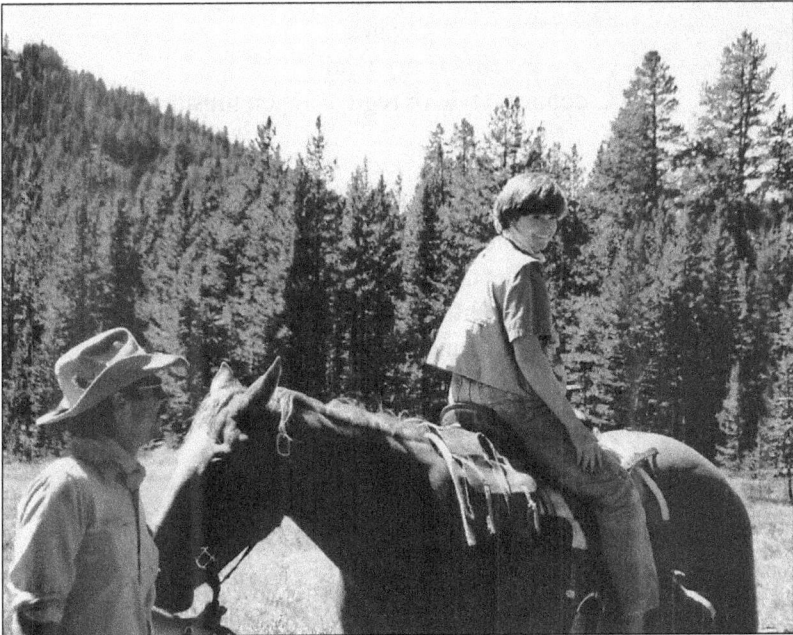

Sausage Lentil Soup

$1/2$ lb. bulk Italian sausage
1 large onion, finely chopped
1 small green pepper, finely chopped
1 small carrot, finely chopped
1 large garlic clove, finely minced
1 bay leaf
2 cans ($14^{1}/2$ oz. each) chicken broth
1 can ($14^{1}/2$ oz. to 16 oz.) whole tomatoes with liquid, coarsely
chopped
1 C water
3/4 C dry lentils
$1/4$ C country-style or regular Dijon mustard

In a Dutch oven, brown sausage.

Drain fat, crumble sausage and return to Dutch oven along with remaining ingredients except mustard.

Simmer 1 hour or until lentils and vegetables are tender.

Stir in the mustard. Remove and discard bay leaf before serving.

POTS, PANS, PONIES & PINES

Barley and Lentil Soup with Swiss chard

1 T olive oil
1^1/2 C chopped onions
1^1/2 C chopped peeled carrots
3 large garlic cloves, minced
2^1/2 t ground cumin
10 C (or more) chicken or vegetable broth
^{2}/3 C pearl barley
1 can (14. 5 oz.) diced tomatoes in juice
^{2}/3 C lentils
4 C coarsely chopped Swiss chard
1 t dried dill

Heat oil in large Dutch oven.
Add onions and carrots.
Sauté until onions are golden brown, about 10 minutes.
Add garlic and stir 1 minute. Mix in cumin, stir.
Add 2/3 C dry lentils, and 10 C broth. Bring to boil.
Reduce heat, partially cover and simmer 25 minutes.
Stir in tomatoes with juice and barley and cover.
Simmer until barley and lentils are tender, about 30 minutes.
Add chard to soup, cover and simmer until chard is tender.
Stir in dill.
Season with salt and pepper.

Cast Iron Cooking on the Trail

POTS, PANS, PONIES & PINES

Salads

Cast Iron Cooking on the Trail

Prewashed bags of salad and salad kits have become an easy way to enjoy a salad at every meal. These bags pack well and save a lot of time. I have used the Caesar Salad kits, coleslaw, spring green mix and prewashed spinach on several trips.

A bag of spinach with mandarin oranges, sliced strawberries and oil and vinegar also makes a nice salad. If I have leftover chicken or steak, I slice it thin, and add it to a bowl of greens with sliced hardboiled eggs and bacon bits.

Place a variety of different bottled salad dressings on the serving table.

Cast Iron Cooking on the Trail

Four Bean Salad

1 can (16 oz.) green beans
1 can (16 oz.) wax beans
1 can (16 oz.) kidney beans)
1 can (16 oz.) garbanzo beans
$1/2$ C chopped celery
$1/2$ C medium onion, sliced into rings
1 small bottle Italian dressing
1 T Sugar

Drain all the beans. Combine all four beans, celery and onion rings in a bowl and season with dressing and sugar. Lightly toss so all ingredients are coated. Cover and chill for several hours if possible.

Turkey Salad

8 C cooked turkey, chopped in one inch pieces
1 can (8 oz.) sliced water chestnuts
2 lbs. seedless grapes
2 C chopped celery
$1^1/2$ C toasted slivered almonds
1 can (8 oz.) pineapple chunks, drained
Add $1/2$ C of mayonnaise and mix well

Broccoli Salad

2 large bunches broccoli
1 medium onion, chopped
1 can (8 oz.) sliced water chestnuts
12 strips cooked bacon, crumbled
$1^1/_2$ C sunflower seeds, salted or unsalted

Dressing:

1 C Miracle Whip
$^1/_2$ C sugar
2 T vinegar

Discard woody part of broccoli stem. Peel remaining stem, chop it and the broccoli heads.

Combine broccoli, onion and water chestnuts.

Mix together all dressing ingredients. Add dressing to salad 2 hours before serving and toss. Chill if possible.

Just before serving, drain off excess dressing and add bacon and sunflower seeds.

Kris's Tomato Salad

Buy one tomato per person, slice into wedges

Toss with the following dressing:

$1/4$ C sour cream
$1/4$ C olive oil
Juice from $1/2$ a lemon
pinch of ground cumin
pinch of oregano
$1/2$ C feta cheese
$1/2$ diced cucumber

Serve chilled if possible

Cast Iron Cooking on the Trail

Cherry Tomato Salad

1 quart cherry tomatoes, halved
$1/4$ C olive oil
3 t vinegar
$1/4$ C minced fresh parsley
1 to 2 t minced fresh basil
1 to 2 t minced fresh oregano
$1/2$ t salt
$1/2$ t sugar
Lettuce leaves, torn into small pieces

Place tomatoes in small bowl.
Combine oil, vinegar, parsley, basil, oregano, salt and sugar.
Mix well.
Pour over tomatoes.
Cover and chill overnight.
Serve on a bed of lettuce.

Chicken Rice Salad

5 C cooked skinless, boneless chicken, chopped
3 C cooked rice (wild rice if available)
$1^1/2$ C small green grapes
$1^1/2$ C diced celery
1 C drained pineapple tidbits
1 C drained mandarin oranges

Dressing:

2 T olive oil
2 T fresh orange juice
2 T vinegar
1 t salt

Combine salad ingredients in large bowl.
Whisk dressing ingredients in small bowl.
Pour over salad and toss.

Greek Salad

1 medium cucumber, peeled and thinly sliced
2 medium tomatoes, chopped
1 can (10 oz.) pitted whole black olives
1 head cauliflower, broken or cut in small pieces
1 head broccoli, broken or cut in small pieces
1 C feta cheese, cubed
1 medium red onion, peeled and thinly sliced

Dressing:

$1/2$ C olive oil
$1/4$ t salt
1 T red wine vinegar
1 T Dijon mustard
$1/8$ t pepper

Put salad ingredients in large bowl.
Whisk dressing ingredients in small bowl.
Pour over salad and toss.

Taco Salad

2 heads iceberg lettuce, torn in small pieces
4 large tomatoes, quartered, then cut each quarter into thirds
2 bunches green onions, sliced thin
1 can (12 oz.) kidney beans (drained)
1 can (10 oz.) sliced olives (drained)
1 lb. cooked lean hamburger, cooled
1 pkg. (10 oz.) shredded cheddar cheese
$1/2$ bag crumbled tortilla chips
1 small bottle (16 oz.) Thousand Island dressing

Mix all salad ingredients.
Pour on the entire bottle of dressing.
Mix well.
Serve with salsa and extra tortilla chips.

Cast Iron Cooking on the Trail

Traditional Tossed Salad

2 bunches red lettuce, washed and torn into small pieces
2 large tomatoes, sliced
1 medium red onion, sliced
1 box (12 oz.) restaurant style croutons

Serve with Anne's salad dressing or dressing of choice

Anne's Salad Dressing

1 small onion, diced
1 C mayonnaise
$1/3$ C olive oil
$1/4$ C ketchup
2 T sugar
2 T vinegar
1 t prepared French's mustard
$1/2$ t salt
$1/2$ t paprika
$1/4$ celery seed
dash pepper

Mix all ingredients in small bowl.
Add 4 oz. crumbled blue cheese. Chill if possible.

Citrus Spinach Salad

1 bag fresh spinach
1 can (10 oz.) mandarin oranges
1 medium red onion, sliced and separated into rings
1 small grapefruit, peeled and sectioned

Dressing:

3 T honey
2 T lime juice
1 t grated lime peel
$1/4$ t ground nutmeg
$1/3$ C olive oil

In large bowl, mix spinach, oranges, onion and grapefruit.
Whisk dressing ingredients in small bowl.
Pour over salad and toss.

Pea Salad

1 (16 oz.) bag frozen peas, thawed and drained
1 C peanuts or cashews
1 C diced celery
$1/4$ C chopped green onions
4 slices cooked crisped bacon, crumbled
Dress with store bought ranch dressing

Cast Iron Cooking on the Trail

Tom's Salad

8 C romaine lettuce, torn in small pieces
1 red apple, cored, sliced thin
$1/3$ C each of blue cheese and candied walnuts, crumbled

Dressing:

2 T olive oil
$1/4$ t grated lemon peel
1 T lemon juice
2 t cider vinegar
2 t minced green onion
1 t honey
salt and pepper to taste

Whisk dressing ingredients in small bowl.
Put apple slices in dressing and let marinate for one hour.
Pour dressing over lettuce and toss.
Sprinkle with cheese and nuts.

Cabbage Plus

$^1/_2$ head cabbage, shredded
1 foil pkg. (10 oz.) cooked tuna or shrimp
2 T sliced almonds
2 T sesame seeds
4 chopped green onions
1 pkg. ramen noodles, uncooked and broken up

Dressing:

$^1/_2$ C olive oil
3 T vinegar
2 T sugar
1 t salt
$^1/_2$ t pepper

Put salad ingredients in large bowl.
Whisk dressing ingredients in small bowl.
Pour dressing over salad. Toss.

Cast Iron Cooking on the Trail

Potato Salad

1 bag small red potatoes
$1/2$ C chopped fresh parsley
$1/2$ C chopped fresh tarragon
$1/2$ C chopped green onion
$1/2$ C chopped celery

Dressing:

1 C mayonnaise
$1/2$ C whole grain Dijon mustard

Boil red potatoes with skins on until tender.
Drain and cut in half.
Mix in other ingredients.
Whisk dressing ingredients in small bowl.
Add to potato salad and mix well.
Salt and pepper to taste.

Cast Iron Cooking on the Trail

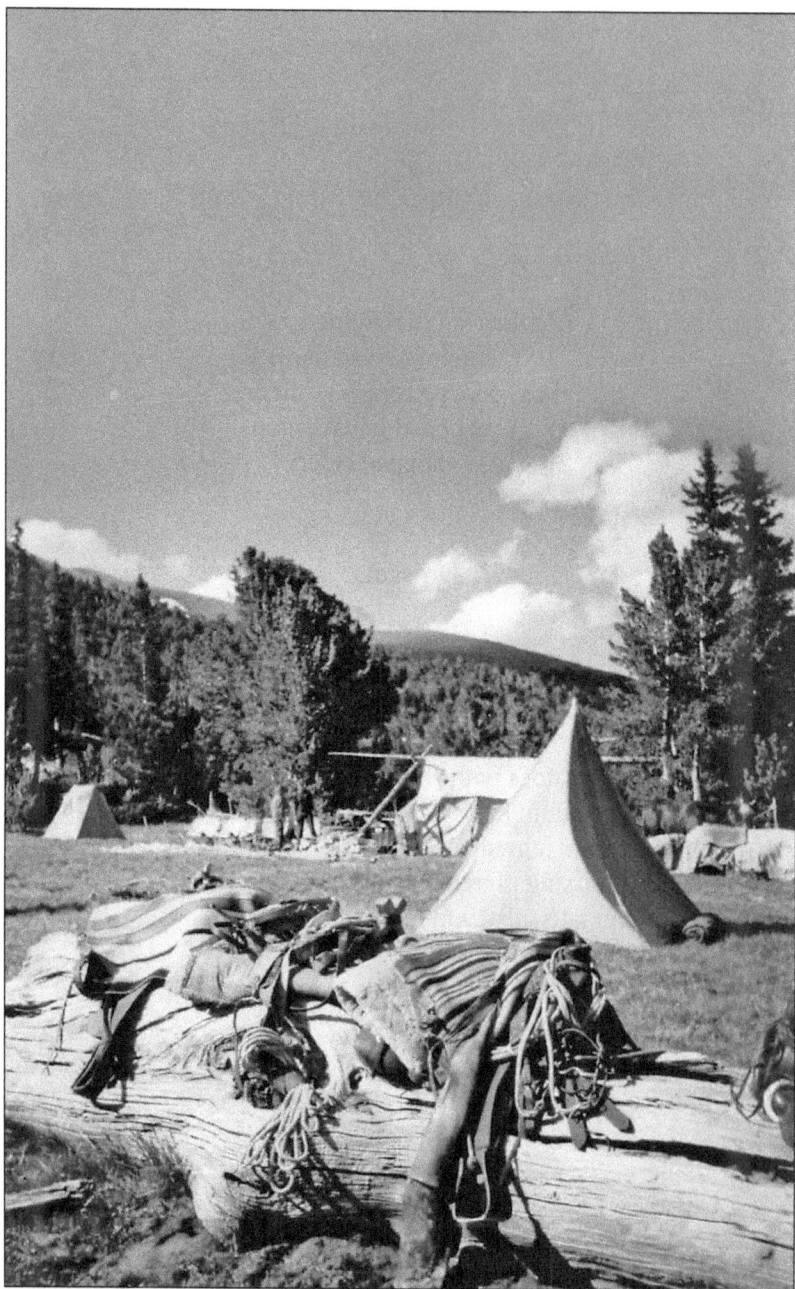

POTS, PANS, PONIES & PINES

Appetizers

Cast Iron Cooking on the Trail

POTS, PANS, PONIES & PINES

Keep it simple on moving days. As soon as possible after getting to camp, set up your serving table. On one end of the table, put out cheese and crackers, peanuts, etc. Something quick and easy that you don't have to cook. Add any bottles of liquor or wine the guests have asked to be packed in. The wranglers will get the beer into the creek to chill.

Make one pitcher of lemonade or flavored fruit drink. Crystal Light or something similar is a good choice. Have another large pitcher of fresh water available. Arrange a washing tub on a camp chair or stump close to the serving table. Put soap and clean towels out each night so folks can wash before eating.

When planning your menus, select appetizers with different tastes and consistencies. Have a crunchy cold dip along with a creamy hot one. Try to gauge appetites. Set out an appropriate amount so guests don't fill up before the main course. Remember, don't buy too many perishable items.

Easy appetizers to pack in:

Assorted cheese and crackers

Pretzels

Potato chips

Pringles—all flavors

Veggies and ranch dip

Tortilla chips with salsa

Peanuts/walnuts/cashews

Crackers—Wheat Thins, Triscuits, club

Cast Iron Cooking on the Trail

POTS, PANS, PONIES & PINES

Pepper Jelly and Cream Cheese

1 tub or 8 oz. cream cheese
Pepper jelly—can buy this in the store
Pour jelly over cream cheese and serve with crackers

Creamy Spinach Dip

Combine:

1 C mayonnaise
1 pkg. Knorr Vegetable recipe mix
1 pkg. (10 oz.) frozen chopped spinach, thawed and squeezed dry
1 container (16 oz.) sour cream

Chill if possible.
Serve with crackers or raw vegetables.

Cast Iron Cooking on the Trail

Crispy Trout

1/4 C butter
1/2 C chopped parsley
1 egg
1/4 C milk
3/4 C dried bread crumbs
1/2 C Swiss cheese
Trout, cleaned

Sprinkle the inside of the trout with salt and pepper.
Combine the softened butter and parsley.
Spread this on the inside of the fish.
Beat egg with milk.
Combine breadcrumbs and cheese.
Dip each fish in egg mixture then roll in crumbs.
Place in skillet with a little oil.
Fry 15 to 20 minutes, flipping once.
Serve on platter with several forks.

Marinated Mushrooms

2 lbs. fresh small mushrooms
1 pkg. Good Seasons Italian dressing mix
$1/2$ C white vinegar
2 T water
$1^1/3$ C oil
$1/2$ t dry ground mustard
Juice of small lemon
1 T Worcestershire sauce
1 t Lawry's seasoning salt
1 t pepper
1 C celery, chopped very fine
1 C onion, chopped very fine
1 C parsley, chopped very fine

Combine all ingredients. Stir once a day.
Ready to serve in 4-5 days.
Should be made ahead at home to take into mountains.
Holds well.

Cast Iron Cooking on the Trail

Mexican Cheese

4 C shredded cheddar cheese
3 chopped green onions
$1/2$ C chopped black olives
1 can (10 oz.) Rotel diced tomatoes with green chilies
1 can (4 oz.) chopped green chilies

Melt cheese in small Dutch oven.
Add all ingredients, stirring occasionally.
Serve with tortilla chips or cauliflower flowerets, zucchini slices or carrot sticks.

Hot Mexican Dip

1 large can (1 lb. 13 oz.) refried beans
2 C sour cream
1 pint salsa
1 C shredded cheese

Layer ingredients in small Dutch Oven. Put in warm pit (no coals underneath) and bank coals on sides and on top.
Cook for 15 minutes, but check after 10 minutes. Cheese should be melted and all ingredients should be warm.

Serve with tortilla chips or crackers

Cheesy Chili Dip

1 pkg. (8 oz.) cream cheese
1 can (14.5 oz.) chili and beans
$1/2$ C shredded cheddar cheese
2 T chopped fresh cilantro

Line small skillet with foil. Spread cream cheese on bottom of skillet and then top with chili and cheddar cheese.

Cover with foil and heat until the cheese melts. You can warm this up on the grate.

Sprinkle cilantro on top before serving.Serve with crackers.

McMarty's Chicken Nuggets

2 boneless, skinless chicken breasts, cut into 1 inch pieces.
2 C Krusteze pancake mix.
2 C oil

Add water and mix pancake batter according to box directions.
Coat chicken pieces with batter
Pour oil in small skillet. Heat on grate until oil is very hot.
Fry until chicken is done, 2 to 4 minutes, to golden brown.
Serve with hot prepared mustard

Cast Iron Cooking on the Trail

Batter Fried Mushrooms

3 to 4 dozen medium to large mushrooms, clean, leave whole

Batter:

2 C all purpose flour
1 T salt
$1^1/2$ t garlic powder
1 t baking powder
$1^1/2$ C beer

Mix all batter ingredients together. Roll mushrooms in batter.
Put 2 C oil in small skillet. Heat oil until hot.
Fry mushrooms until batter turns a golden brown. Drain on paper towel and season with salt and pepper

Homemade Salsa
(Worth the effort if you have the time!)

$1/2$ C chopped green onions
1 small can (4 oz.)diced green chilies
$1/4$ C chopped fresh cilantro
3 ripe tomatoes, diced
$1/4$ C lemon juice
1 red bell pepper, diced
1 T rice vinegar
$1/2$ t peanut oil
$1/4$ t Tabasco sauce

Mix all ingredients. Serve with tortilla chips

Crab Melts

1 pkg. English muffins (8 count)
2 cans (8 oz.) crab meat
1 small pkg. (1 lb) Velveeta cheese
1 stick butter, softened
1/2 small onion, grated
1 of the following seasonings: onion, garlic or celery salt.

Cream butter and cheese.
Drain crab meat.
Add crab meat, onion, and seasoning to cheese mixture.
Spread on bottom half of muffin, top with other half.
Grill on skillet. Turn over until both sides of English Muffin are brown and cheese/crab is melted and hot.
Cut into quarters.

Variation:

Add 1/2 C green or red pepper, finely chopped

These can also be made for lunch on a layover day.

Cast Iron Cooking on the Trail

Crab Ball

1 can (8 oz.) crab meat
1 pkg. (8 oz.) cream cheese
2 T chopped chives
$^1/_4$ t garlic powder
$^1/_4$ t salt
$^1/_2$ C chopped pecans.

Drain crab meat.
Blend softened cream cheese, chives, garlic powder and salt.
Fold in crabmeat. Shape into ball.
Serve with crackers or fresh vegetables.

Hot Artichoke Dip

1 large can (14.5 oz.) artichoke hearts, drained and chopped
$^1/_2$ C mayonnaise
1 small can (4 oz.) chopped green chilies
$^1/_2$ C Parmesan cheese.

Mix all ingredients. Line small skillet with foil. Put artichoke mixture in skillet. Cover with foil. Heat through on top of grate. Serve with small slices of rye bread, or crackers.

Nacho Dip

1 lb. Velveeta cheese
1 large onion, chopped
1 can (10 oz.) Rotel diced tomatoes and green chilies
1 small can (4 oz.) fire roasted diced green chilies

Melt cheese over low heat. Sauté onions in $1/3$ C water. Add tomatoes and chilies. When all liquid is gone, add to cheese. Serve with tortilla chips.

Curry Paté

2 pkg. (3 oz. each) softened cream cheese
1 C shredded sharp cheddar cheese
3 T dry sherry
$1/2$ t curry powder
$1/4$ t salt
1 jar (8 oz.) mango chutney
$1/3$ C green onions, finely chopped

Beat together first five ingredients.
Spread on a plate; chill until firm.
At serving time, bring to room temperature and spread with chutney and finely chopped green onions.
Serve with crackers or pita strips.

Cast Iron Cooking on the Trail

Warm Black Bean Dip

1 small onion, minced
2 minced garlic cloves
1 T vegetable oil
1 can (15 oz.) black beans, rinsed and drained
$1/2$ C diced fresh tomato
$1/3$ C picante sauce
$1/2$ t ground cumin
$1/2$ t chili powder
$1/4$ C Mexican cheese blend or cheddar cheese
1 T lime juice
cilantro for garnish

In a medium skillet, sauté onion and garlic in oil until tender.
Add the beans; mash gently.
Stir in the tomato, picante sauce, cumin and chili powder.
Cook and stir just until heated through.
You can use the grate to cook on.
Remove from the heat, stir in cheese, cilantro and lime juice.
Serve warm with chips.

Bacon Tomato Spread

1 pkg. (8 ounces) cream cheese
2 t ground mustard
$1/2$ t celery seed
1 medium tomato, peeled, seeded and finely chopped
$1/4$ C green pepper
8 bacon strips, cooked and diced

In a mixing bowl, beat cream cheese, mustard and celery seed until blended.
Stir in tomato and green pepper. Cover and chill.
Cook bacon until crisp.Crumble.
Stir bacon into spread before serving.
Serve with crackers or raw vegetables.

Sour Cream Salsa Dip

1 large sour cream (32 oz.)
$1/2$ C chunky hot salsa
1 pkg. Knorr vegetable soup mix

Mix well. Serve with tortilla chips or veggies.

Cast Iron Cooking on the Trail

Yummy Fruit Dip

8 oz. cream cheese
Small jar (8 oz.) Jet Puff marshmallow cream
Lemon zest
$1^1/_2$ pints whole strawberries, washed

Mix well.
Serve with whole strawberries

Red Pepper Crackers

1 large box saltine crackers
$1^1/_2$ C peanut oil
3 T dry ranch dressing mix
2 T red pepper flakes

Mix in large Ziplock bag. Be sure to shake up the bag to get all the ingredients on the crackers.
Serve with pepper jack cheese and your favorite adult beverage.

Stuffed Jalapeño Peppers

8 jalapeño peppers
1 pkg. (8 oz.) cream cheese
1 C cheddar cheese
dash of garlic powder
dash of salt
$1/2$ lb. partially cooked hickory smoked bacon

Cut jalapeños in half length-wise and remove seeds.
Combine cream cheese and Cheddar cheese.
Add spices and mix well.
Stuff peppers generously with cheese mixture.
Wrap each pepper with a slice of bacon.
Place in skillet on grate, cover with foil.
Cook for 10 to 12 minutes until cheese is melted
and bubbly or until bacon is crisp.

Pork chops with mushroom cream gravy.

POTS, PANS, PONIES & PINES

Skillet Meals

Cast Iron Cooking on the Trail

POTS, PANS, PONIES & PINES

Use large cast iron skillet or medium Dutch Oven.

      Skillet menus are useful when there is a fire ban in place in the back country and you have to use propane for fuel.

      Most people are familiar with Coleman Stoves, but when you are using one you don't have the advantage of burning wood and using coals. This is the perfect time for Skillet Meals.

      I have identified other recipes in other sections that could also be used for Skillet menus by putting an asterisk beside the recipe. Think of it as cooking on your stove top instead of in your oven.

Cast Iron Cooking on the Trail

Sweet and Sour Pork

$1^1/_2$ lbs. pork loin cut into 1 inch cubes
1 C boiling water
1 chicken bouillon cube
1 t salt
1 large can (1 lb. 12 oz.) pineapple chunks
$1/_4$ C brown sugar
2 T cornstarch
$1/_4$ C vinegar
1 T soy sauce
1 green pepper, cut in strips
1 onion, sliced
Rice

Brown pork in shortening. Add 1 C boiling water, chicken bouillon cube and salt. Stir until bouillon cube is dissolved.

Mix well and cover. Simmer on low until tender, about 1 hour.

Remove pork and broth from skillet. Set aside.

Drain pineapple chunks, reserving the syrup. Set them aside.

Put reserved pineapple syrup, vinegar, soy sauce and salt in large skillet. Combine brown sugar and cornstarch and add to mixture.

Cook and stir over medium heat until thick and bubbly.

Remove from heat.

Add pork to mixture and mix well. Stir in pineapple chunks, green pepper and onion.

Cook over low heat 2 to 3 minutes or until vegetables are tender crisp. Serve over rice.

Cast Iron Cooking on the Trail

Chili

1$\frac{1}{2}$ lbs. ground beef
1 large diced yellow onion
2 chopped cloves of garlic
2 large cans (1 lb. 12 oz.) tomato sauce
2 large cans (1 lb. 12 oz.) Italian stewed tomatoes
1 large can (1 lb. 12 oz.) dark red kidney beans, drained
1 small can (4 oz.) diced green chilies (hot, fire roasted)
$\frac{1}{2}$ C diced red bell pepper
3 T chili powder (or season to taste)
1 t salt
1 t dried oregano
1 t pepper
$\frac{1}{2}$ C brown sugar

Brown meat, onion and garlic in skillet. Drain. Add other ingredients, cover and simmer on low for an hour or more. Serve with Mexican style cornbread.

To make a hotter chili, you can add cayenne pepper and red pepper flakes. Add a little at a time and then taste.

Tim's Chili

4-5 cloves of garlic chopped
3 lbs. stew meat
2 lbs. lean hamburger
2 t cumin
2 to 4 T chili powder (or season to taste)
2 onions-chopped
$1/2$ t ground cinnamon
2 cans (15 oz.) each:
pinto, great northern, kidney and black beans
Total: 8 cans of beans
4 cans (15 oz.) whole tomatoes
1 T white sugar

Sauté garlic and onions in small amount of olive oil until onions are translucent.

Brown stew meat and hamburger separately and add to large Dutch oven.

Add garlic, onions and remaining ingredients. Simmer on low for a couple of hours.

Cast Iron Cooking on the Trail

Pork Green Chili

2 T vegetable oil
1^1/2 lbs. cubed pork (pork loin)
2 T all-purpose flour
1 can (4 oz.) diced green chilies, drained
^{1}/2 can (3.5 oz.) chopped jalapeno peppers
^{1}/2 medium onion, chopped
1 can (14.5 oz.) Fire Roasted diced tomatoes
3 C water
onion salt to taste
garlic salt to taste
salt and pepper to taste

Sprinkle flour over cubed pork.
Heat oil in a large cast iron skillet over medium-high heat.
Stir in cubed pork and cook until brown and cooked through, about 15 minutes.
Remove skillet and allow to cool.

Add chilies, jalapeno peppers and onions. Stir in tomatoes and water. Season to taste with onion salt, garlic salt, and salt and pepper

Return skillet to medium heat. Bring to a simmer, cover and cook 30 minutes, stirring occasionally.
Remove cover and cook 10 minutes more. Serve with warm tortillas, rice or beans.

Calico Beans

$1/2$ lb. bacon, chopped
1 can (15 oz.) each:
pork and beans, black beans, kidney beans, and northern beans
(Total:4 cans beans)
1 lb. ground beef
1 C chopped onion
1 chopped clove of garlic
$1/2$ C ketchup
$1/4$ C brown sugar
$1/4$ C white sugar
1 t prepared mustard
1 t salt
2 t vinegar
1-2 T Liquid Smoke to taste

Cook bacon in large skillet or Dutch oven.
Remove with slotted spoon to drain.
Add beef, garlic and onion to bacon grease.
Cook until onion is tender. Drain off excess grease.
Drain cans of beans and add to meat mixture.
Add remaining ingredients.
Simmer for 30-40 minutes.

Macaroni and Cheese

1 lb. pkg. macaroni shells
$1/2$ to $3/4$ pkg. Velveeta (2 lb.) cheese, cut into 1 inch pieces
1 stick butter
salt and pepper to taste

Boil macaroni in large pot with boiling water with a little salt
Cook until *el dente*.
Allow extra time to cook pasta in the mountains. It takes longer to get a rolling boil.
Drain the pasta and put in medium Dutch oven.
Add the cheese and butter.
You can add a little milk or cream if you think it is too thick.

Meatball Sandwiches

Meatballs:

1 lb. hamburger	1 lb. bulk Italian sausage

2 eggs, lightly beaten
$1/2$ C dry breadcrumbs

3 T milk	1 t dried basil
3/4 t salt	3 T olive oil

Sauce:

$1/2$ lb. bulk Italian sausage

$1/2$ C chopped onion	$1/2$ C chopped green pepper
$1^1/2$ C stewed tomatoes	1 can (6 oz.) tomato paste

2 cans (15 oz each) tomato sauce

2 t white sugar	1 t garlic powder
$1/2$ t dried oregano	$1/2$ t dried basil

8 slices Mozarella cheese
1 small jar Parmesan cheese

In a Dutch oven, cook sausage, onion and green pepper until the sausage is browned and the vegetables are tender. Drain.
Add the remaining sauce ingredients; bring to a boil.
Cover and simmer.
Meanwhile, in a bowl, combine first 7 meatball ingredients. Shape into 1 inch balls. Brown in oil. Drain.

To assemble:

Slice big crusty rolls so you have a top and bottom. Place meatballs on bottom, spoon sauce on top, place a slice of mozzarella cheese or sprinkle with Parmesan on top of that and cover with top of roll. Wrap in foil and keep warm on grate.

Can be used as a layover lunch.

Cast Iron Cooking on the Trail

Barbecue Meatballs

Meatballs:

1^1/2 lbs. ground beef
3/4 C rolled oats
2 eggs, slightly beaten
^{1}/2 C finely chopped onion
^{1}/2 C milk
1 t salt
^{1}/4 t pepper
1 t Worcestershire sauce

Combine & mix well.
Roll into 1 to 1^1/2 inch meatballs.
Put 2 T oil in skillet. Cook until brown on the outside and cooked through on the inside.

Sauce:

^{1}/2 C brown sugar
^{1}/4 C vinegar
1 t French's yellow mustard
1 C BBQ sauce—choose your favorite brand
1 t Worcestershire sauce

Combine all ingredients and mix well. Heat in small saucepan until smooth and bubbly. Pour over meatballs. Cover and simmer on for 30-35 minutes. Season to taste

Makes a great appetizer.

POTS, PANS, PONIES & PINES

Meat Sauce with Penne pasta

Sauce:

6 T olive oil
1 yellow onion, chopped
1 small carrot, peeled and chopped
1 small celery stalk, chopped
1 T dried chopped parsley
$1/2$ lb. ground beef
$1/2$ C dry red wine
1 large can (lb. 12 oz.) diced tomatoes
salt and pepper to taste
$1/2$ C beef broth

Pasta:

1 lb. penne pasta
2 T salt   2 T butter
1 C grated Parmesan cheese, or as much as you want

Warm olive oil in Dutch oven over medium heat.

Add onion, carrot and celery stirring often, until vegetables are soft, about 10 minutes.

Stir in parsley and after one minute add the meat.

When the meat is brown, add the wine to the pan and stir until the alcohol is evaporated and the liquid is reduced.

Reduce the heat, add the tomatoes, season and stir frequently until the liquid has been reduced.

Add broth and simmer sauce for about 40 minutes.

Cook pasta until tender.

Transfer the sauce to large skillet.

Add the cooked pasta and butter and stir until the sauce is almost completely absorbed, about 7 minutes. Sprinkle with Parmesan and then place cheese on the serving table

Cast Iron Cooking on the Trail

Penne Pasta with Gorgonzola Cheese Sauce

$1^1/_2$ lbs. asparagus, cut into $1^1/_2$ inch pieces
2 C (about 8 oz.) crumbled gorgonzola cheese
3/4 C whipping cream
1 pkg. (16 oz.) penne or rigatoni
2 T lemon juice

Cook asparagus pieces in large pot of boiling salted water until crisp-tender (about 2 minutes). Rinse in cold water and set aside.

Place crumbled gorgonzola cheese and whipping cream in small Dutch over. Cook over very low heat until cheese melts and mixtures are almost smooth (about 4 minutes).

Meanwhile cook pasta in large pot with boiling salted water until tender, stirring occasionally.

Drain well.

Return pasta to large skillet.

Add gorgonzola sauce to pasta. Toss to coat.

Mix in asparagus.

Drizzle lemon juice over.

Season to taste with salt and pepper.

Can serve with Parmesan cheese if desired.

Spaghetti with Meatballs

Sauce:

1 small onion, chopped
3 cloves garlic
1 can (4 oz.) mushrooms
1 T olive oil
2 large cans (1 lb. 12 oz.) tomato sauce
2 large cans (1 lb. 12 oz.) whole tomatoes
1 can (6 oz.) tomato paste
3 "paste" cans of water
1 T chopped dried parsley
2 bay leaves
1 t dried oregano
1 t dried basil
1 t sugar
2 T Parmesan cheese
salt and pepper

Meatballs:

1 lb. lean hamburger
1 lb. sweet or hot Italian sausage
1/2 large onion, finely diced
3 eggs
1 T chopped parsley
1/2 C breadcrumbs
salt and pepper

Spaghetti noodles – Family pkg.

To make the Sauce:

Cast Iron Cooking on the Trail

In large Dutch oven, sauté onions and garlic in small amount of olive oil until onions are translucent.

Add remaining ingredients and cook slowly for one hour or longer.

Mix meatball ingredients together, form into $1^1/2$ inch balls and brown thoroughly in lightly oiled skillet.

Add to the spaghetti sauce mixture and allow meatballs to simmer in the sauce for at least an hour.

I fill a clean aluminum water bucket with water and a small amount of olive oil to cook the spaghetti in.

*I plan 2-3 ounces of dry spaghetti per person.

Allow a little extra time in the mountains for the water to boil.

Drain the spaghetti. Serve meatballs and sauce over spaghetti.

Place Parmesan cheese on serving table.

Kate's Pasta

1 pkg. (16 oz.) bowtie pasta
(Cook it *al dente* because dressing will soften it)
1 red onion, chopped
1 bunch green onions, chopped
1 can (10.5 oz.) pitted black olives
$1/2$ lb. Bacon, chopped
peppers of all colors: red, yellow, orange, anything but green
leftover chicken or pork (as much as you want)
handful of grape tomatoes
$1/2$ pkg. frozen peas, thawed

Mix in 1 can (14.5 oz.) of diced tomatoes with basil/garlic

Dressing:

Olive oil, red wine vinegar, 1 t Dijon mustard and salt and pepper
Or use Wishbone Italian

Top with Parmesan cheese or goat cheese.
Toast pine nuts or walnuts and sprinkle along with fresh basil over each serving.

Cast Iron Cooking on the Trail

Alfredo Sauce

1 pint cream or half and half
1 stick butter
2 T cream cheese
$^1/_2$ -3/4 C Parmesan cheese
1 T garlic powder

Melt butter and cream cheese. Add garlic.
Simmer for 15 minutes.
Add Parmesan cheese and pour over any pasta that you like

Optional: Add sautéed broccoli and mushrooms.

John Ben Getty Casserole

$1^1/_2$ lbs. hamburger meat
2 chopped onions   1 green pepper chopped
16 oz. small shell pasta
1 can (14.5 oz.) peas, drained
1 can (10.5 oz.) mushroom soup
1 can (10.5 oz.) tomato soup
$^1/_2$ C grated cheddar cheese
salt and pepper to taste

Brown hamburger, onions, and peppers in skillet. Add peas, mushroom soup and tomato soup plus seasoning.
Cook pasta according to package directions.
Alternate this mixture with pasta in medium skillet. Place tin foil on top and simmer on low heat until cooked through.
Sprinkle with grated cheese before serving.

Blackened Chicken

4 chicken breast halves, cut in half to make thinner.
(These are sometimes called cutlets.

Cajun Mixture:

2 $^1/_2$ T paprika
2 T garlic powder
1 T salt
1 T onion powder
1 T dried thyme
2 T ground red pepper
1 T black pepper

Rub chicken breasts with Cajun mixture.
Put 3 T oil in medium skillet. Add chicken and cook 7 minutes on each side or until done.
Serve whole or cut across grain into thin slices.

The thin slices are really yummy in a green salad.

Chicken and Broccoli

2 pkgs.frozen broccoli
1 pint sour cream
1 pkg. dry onion soup mix
2 C diced cooked chicken (or turkey)
1 C heavy cream
1 T cheddar or Parmesan cheese, grated

Cook broccoli as directed on package; drain.
Mix together sour cream and dry soup mix.
Place the drained broccoli on bottom of medium skillet.
Cover with half the sour cream mixture.
Add half the heavy cream.
Put the diced chicken on top
Pour remaining combined cream mixture over all.
Simmer for 20 minutes on very low heat.
Sprinkle cheese on top.

POTS, PANS, PONIES & PINES

Baked Chicken with Rice

1 can (10.5 oz.) cream of celery *or* tomato condensed soup
(for your preferred flavor)
diluted with $1/2$ soup can of water
1 can (10.5 oz.) condensed onion soup
diluted with $1/2$ soup can of water
$1/2$ C uncooked wild rice
1 C uncooked regular white rice
6-8 pieces of chicken
(assortment of breasts, thighs or drumsticks)
1 can (10.5 oz.) undiluted mushroom soup

Brown chicken in small amount of oil. Set aside.

Mix celery or tomato soup with onion soup and water.
Combine wild and white rice and blend with soup mixture.
Spoon into large skillet. Arrange chicken pieces on top.
Spread chicken with mushroom soup.
Bake, covered, for 1 hour and 45 minutes or until chicken and rice is done. Check and stir often.
If you can, cover the skillet with foil, put foiled Jackson Grill on top and add a few coals for the last 30 minutes.

Cast Iron Cooking on the Trail

Wild Rice Hamburger Casserole

Combine in skillet:
$^1/_2$ C washed wild rice
$1^1/_4$ C water (or beef bouillon)
Cover with lid and place over medium heat. Cook from 30-45
minutes or until rice is fluffy. Remove rice from skillet. Set aside.

Brown:
1 lb. hamburger

Mix in:
1 C chopped celery
$^1/_2$ C chopped onion
$^1/_4$ C chopped or sliced carrots
$^1/_4$ C sliced parsnips (or any vegetable combination)

Add:
Cooked wild rice (plus remaining liquid)
1 can (10.5 oz.) cream of mushroom soup
$1^1/_2$ C water
$^1/_2$ C uncooked white rice
salt and pepper to taste

Stir until blended. Cover with lid and continue cooking (20 to
30 minutes) or until rice is fluffy and liquids are absorbed.

Mexican Chicken Casserole

4 chicken breasts- cooked and sliced thin
2 cans (10.5 oz.) cream of chicken soup
1 C milk
1 medium chopped onion
2 small cans (4 oz.) diced green chilies
1 pkg. (30 count) corn tortillas
1 pkg. (16 oz.) grated cheddar and Monterey Jack cheese

Combine soup with milk. Add chopped onions and chilies.

Start with a layer of chicken.
Then layer with 13-14 fried corn tortillas.
Layer $1/2$ the grated cheese.
Pour half of the soup mixture and repeat layers.
Heat over low heat for 30 minutes or until bubbly.
Serve with diced tomatoes and lettuce.

Cast Iron Cooking on the Trail

Main Dishes

Cast Iron Cooking on the Trail

POTS, PANS, PONIES & PINES

Occasionally in dry years, the Forest Service or the Park Service will put a fire ban on forest and wilderness areas. This means that wood cannot be burned. You cannot have a fire for warmth or for meal preparation. There are no exceptions.

In this section, I list main dishes which I feel can be the core of the dinner meal. Some of them would work well for skillet dishes. Skillet dishes can be cooked on a grate with coals underneath or over a propane burner that would be packed into the mountains.

Propane burners, which most people recognize as Coleman Stoves, are good stoves. Your food will take longer to cook and you will have to check it more often. The size of the propane burners dictates the size of the cooking pans.

I have put an asterisk by recipes in this section that could also be used for skillet meals.

If there is plenty of wood and there is not a fire ban, using a Dutch oven is the way to go. You can put the main dish in the Dutch oven, put the oven in your cooking pit with coals on the bottom, coals around the side and on the top.

Check it halfway through the cooking time, give it a stir, rotate the oven and leave it. It will be cooked perfectly and will stay warm while you're serving.

Cast Iron Cooking on the Trail

Hint: If the recipe calls for cheese, line the Dutch oven with foil to make for an easier cleanup. Also cover the top of the oven, before putting the lid on. This prevents the ash from getting into the food when sweeping the coals away from the oven. And finally, cover the Jackson Grill with foil if you plan to use it for a skillet lid.

Chicken Mar-Sal (Marsala)

8-10 boneless, skinless chicken breasts
$^1/2$ pint sour cream
1 C cooking sherry
1 can (10.5 oz.) golden cream of mushroom soup
1 can (4.5 oz.) mushrooms
paprika

Coat chicken in paprika.

Mix remaining ingredients together and pour over chicken.

Put in Dutch oven and place in cool pit (no coals on bottom), putting medium coals around sides and on top.

Cook for approximately 1 to $1^1/2$ hours.

Check often and rotate the oven. You might also want to move the chicken around in the oven.Put chicken that is on the bottom, on top and reverse the layers.

When cooking chicken breasts, I cut into each breast to make certain it's done.

Serve with instant rice.

Leftover chicken can be sliced in a salad the next day or can be used in sandwiches.

Cast Iron Cooking on the Trail

Chicken Parmesan*

8-10 boneless, skinless chicken breasts
3 eggs, whisked
1 C milk
Italian breadcrumbs
tomato sauce (1 lb. 12 oz.)
Mozzarella cheese, 1 slice for each piece of chicken
Parmesan cheese to sprinkle on top of each piece of chicken

Cut chicken breasts in half to make thinner.
Put between wax paper and pound chicken breast to $1/4$ inch thickness. You can use a rolling pin or tent stake mallet.
Dredge in egg and milk, and repeat.
Coat with seasoned flour or breadcrumbs.
Fry in large skillet until done.
Top with tomato sauce, mozzarella cheese and Parmesan cheese.
Cover with foil. Put foiled Jackson Grill on top.
Place in cool pit (no coals on bottom)
Cover with medium coals on sides and on top of Jackson Grill.
Cook until cheese is melted, 5 to 10 minutes.
Serve with spaghetti noodles and additional spaghetti sauce.
Have additional Parmesan cheese on serving table.

*There are several spaghetti sauce recipes to choose from in the skillet menu section, or buy your favorite brand at the store.

POTS, PANS, PONIES & PINES

Bethany's Picante Chicken*

6 skinless, boneless chicken breasts

Sauce:

1 small jar (12 oz.) picante salsa
$^1/_2$ C brown sugar
3 T French's yellow mustard
1 can (4 oz.) diced green chilies.
(Use fire roasted if you want it a little hotter.)

If using chicken with skin, brown breasts first.
Combine sauce ingredients and pour over chicken.
Cover with foil. Top with foiled Jackson Grill.
Place in fire pit (cool).
Put coals around side and on top.
Cook for 35 minutes over medium heat.
Serve with rice.

Cast Iron Cooking on the Trail

Chicken Enchiladas Casserole

1 medium chopped onion
2 to 3 T butter
1 can ($10^1/2$ oz.) cream of chicken soup
1 can ($10^1/2$ oz.) chicken broth
1 to $1^1/2$ small cans (4 oz.) chopped green chilies
3 boneless, skinless cooked chicken breasts, chopped or shredded
1 pkg. corn tortillas (1 lb. 9 oz.) broken into pieces
1 lb. cheddar cheese, grated

Brown onion in butter. Combine onion with soup, chicken broth, green chilies and pieces of chicken. Mix well and heat.
Place in medium Dutch oven in the following order:
Chicken mixture
Layer of tortillas
Chicken mixture.
Repeat and finish with cheese.
Cover with foil, put Dutch oven lid on and place in cool pit (no coals on the bottom) Bank medium coals on sides and top.
Bake for 30 minutes until bubbly, but check after 15.

Turkey Enchiladas*

3 C picante salsa
$^1/_2$ C water
3 T chili powder
4 C chopped turkey, small pieces
2 C (8 ounces) grated cheddar cheese
$^1/_2$ C cilantro chopped
6 (10 inch) flour tortillas

Combine salsa, water and chili powder.
Pour half the mixture into medium Dutch oven.
Set remainder aside
Combine turkey, 1 C of the cheese, and cilantro.
Lay a tortilla on work surface. Place 1 C turkey mixture in center of tortilla, roll closed and place seam side down in Dutch oven. Repeat with remaining tortillas. If tortillas seem a little dry, warm them up and they will roll easier.
Pour reserved sauce over enchiladas.
Cover with foil and top with Dutch oven lid. Place in cool cook pit (no coals on the bottom). Bank medium coals on sides and top of Dutch oven.
Bake about 30 minutes or until bubbly. Check and rotate at 15.
Uncover and sprinkle remaining cheese on top.
Continue baking until cheese melts.

Cast Iron Cooking on the Trail

Mexican-style Chicken Roll-up

8 boneless, skinless chicken breasts
4 cans (7 oz.) whole green chilies, halved and seedless
3 oz. Monterey Jack cheese, cut into 8 strips
3/4 C fine, dry breadcrumbs
1 T chili powder
$1^1/_2$ t ground cumin
$^1/_4$ t salt
$^1/_4$ t garlic powder
$^1/_4$ C skim milk
2 T butter

4 C shredded lettuce
$^1/_2$ C picante sauce
$^1/_4$ C sour cream

Trim excess fat from chicken.

Place each chicken breast half between 2 sheets of waxed paper; flatten to $^1/_4$ inch thickness using a rolling pin or a meat mallet.

Place a green chili half and one strip of cheese in center of each chicken breast half. Roll up lengthwise, tucking edges under. Secure with tooth picks.

Combine breadcrumbs, chili powder, cumin, salt and garlic powder. Roll the chicken breast in this mixture.

Melt 2 T butter in medium Dutch oven.

Put chicken rolls in Dutch oven and put oven in a cool pit, (no coals on the bottom). Bank coals on sides and top.

Check chicken after 15 minutes. Rearrange breasts, putting some on top, some on bottom and then reverse the layers.

Continue to cook for another 15 minutes. Cut into each piece of chicken to make sure it's done.

Place each chicken roll on $^1/_2$ C shredded lettuce. Top with picante sauce and $1^1/_2$ t sour cream.

Best Barbecue Chicken

1 family size pkg. of cut up chicken:
breasts, thighs, wings and legs

Brine:

$1/2$ bag (2 lb.) brown sugar
2 C salt
2 T thyme

Sauce:

2 small onions, chopped
1 C ketchup
$2^1/2$ C red wine vinegar
$1/2$ C molasses
1 T paprika
1 t yellow French's mustard

Soak chicken in brine water for 2-3 hours, pat dry.

Grill each piece of chicken for 10 minutes on each side.
Put in Dutch oven with barbecue sauce.
Bury in cool pit (no coals on bottom). Bank medium coals on sides and top. Bake for 30 minutes or until chicken is cooked through.
Check half way through the cooking time. Rearrange pieces of chicken. Adjust coals if necessary and rotate the oven.

Cast Iron Cooking on the Trail

Grilled Chicken

2 whole chickens
(split in half so you would have a leg on each half)

Barbecue sauce: Use your favorite brand or favorite recipe
Marinate chicken in barbecue sauce for 2-3 hours
Grill slowly on grate, over low to medium coals.
Turn often, cook for $1^1/2$ to 2 hours.

Lazy Day Chicken*

6 boneless chicken breasts (whole or in strips)
$^1/2$ C melted butter
1 C crushed stuffing mix (Herb Flavor)
$^1/2$ C Parmesan cheese
1 T dried, crushed parsley

Cut chicken in half to make thinner. (You can cut again in strips if you want.)

Mix the dry ingredients together. Dip chicken in melted butter and roll in dry mixture.

Put in large skillet, cover with foil and top with foiled Jackson Grill. Put in medium pit with a few coals on bottom around sides and on top.

Check after 10 minutes and rotate the skillet. Continue to cook for another 10 minutes.

Variation: Instead of dipping chicken in butter; spread $^1/2$ C mayonnaise and $^1/2$ C Parmesan cheese on chicken, then sprinkle with Italian seasoned breadcrumbs.

You can also use different seasoned breadcrumbs instead of stuffing mix.

Chicken and Dumplings

1 large onion, chopped
6 C chicken broth
2 T poultry seasoning
1 C chopped carrots   1 C chopped celery
Chicken gravy, small jar, (8 oz.)
4 skinless, boneless, chicken breasts-cubed (2 inch pieces)
1 roll Pillsbury refrigerated biscuits, each biscuit cut into quarters

Sauté onion and chicken in medium Dutch oven.
Cook over slow heat for 10 min.
Remove chicken and onions.
Add chicken broth to Dutch oven, season with poultry seasoning.
Add chopped carrots and celery and simmer until tender.
Add cooked chicken and onions.
Add chicken gravy to thicken.
Bring to a boil and drop in biscuits one at a time.

Or make Dumplings from scratch:
2 C flour
2 T baking powder
$1/2$ t salt
2 T shortening
3/4 C hot water or broth

Mix dry ingredients.
Work in shortening and add liquid gradually.
Toss on floured board and roll $1/4$ inch thick.
Cut into small pieces and drop into boiling liquid.
Cook until tender, about 15 minutes.

Chicken Pot Pie

1 pkg. (10 oz.) frozen peas
1 pkg. (10 oz.) frozen carrots
$1/3$ C butter
$1/3$ C flour
$1/3$ C chopped onion
salt and pepper to taste
2 C chicken broth
$2/3$ C milk
3 C cut up cooked chicken or turkey (1 inch pieces)

Crust:

2 C flour
1 C lard or Crisco
4 T cold water
pinch of salt

Rinse frozen peas and carrots and drain.
Let thaw if still frozen.
Make white sauce: melt butter, add the flour, salt and pepper.
Stir in broth and milk. Heat to boiling, stirring constantly.
Boil and stir 1 minute.
Stir in chicken, peas and carrots. Remove from heat.

Make Crust:

Take $2/3$ of the raw pastry dough, roll out and put into large skillet. Pour chicken mixture into crust. Roll out remaining pastry dough and place on top. Cut venting holes. Seal edges.

Place skillet in cool pit, (no coals on the bottom). Cover with foil, top with foiled Jackson Grill, bank coals around sides and put coals on top. Cook over medium heat but rotate skillet often.

Bake until crust is brown and filling is hot; about 40 minutes.

Easy Chicken Cordon Bleu

6 boneless, skinless chicken breasts
1 can (5 oz.) plain breadcrumbs
6 slices of prosciutto ham
6 slices of Swiss cheese
1 pint sour cream
2 cans (10.5 oz.) cream of chicken soup

Cut chicken in half to make thinner then place chicken breasts between two sheets of wax paper and pound until $1/2$ inch thick.

Dip chicken in breadcrumbs.

Place 1 slice of ham and 1 slice of cheese on each chicken breast.

Roll up and secure with a toothpick.

Buy the longer sandwich toothpicks.

Place in greased Dutch oven.

Mix sour cream and soup, then pour over chicken.

Put lid on and bury in cool pit (no coals on the bottom.)

Bank sides and top with medium coals.

Cook for 45 minutes, but check at 20.

Rotate Dutch oven often.

Cast Iron Cooking on the Trail

Cornish Game Hens

6-8 Cornish game hens, cleaned out and washed
3 T oil
4-6 carrots, cut into 2 inch chunks
3-4 potatoes, cut into 2 inch chunks
2 onions, quartered
salt and pepper

Brown birds in skillet in oil. Place in large Dutch oven.
Fill with water to cover.
Add carrots, potatoes and onions. Make sure water covers game hens and vegetables.
Bury Dutch oven in hot coals. Place hot coals under the bottom of the Dutch oven, then bank coals around sides and place on top. Use a lot of coals.
Cook for 4 hours. Season to taste.

This recipe is great if you want to leave the camp to go fishing, hike, etc. You don't have to check it, but make sure someone always stays in camp when you have a fire burning.

Stuffed Pork Loin

Plan on about $1/2$ lb. per person
2 pork loins: 6 lbs.
prepared Stove Top stuffing per box instructions

Brown loin in 2 T lard or oil. Slice loins lengthwise but leave one side attached. Put stuffing in the middle, close and put more stuffing over the top.

Place Dutch oven in warm pit with hot coals underneath. Bank coals around sides and put some on top.

Check after 15 minutes in case you need to adjust the heat. Cook for about 40 minutes.

Take Dutch oven out of the pit when meat is still a little underdone. The meat will continue to cook.

For large Dutch oven, plan to check meat often and continue cooking until meat is a light pink color. Slice meat and serve with stuffing. Pork sandwiches make a delicious lunch the next day, so I usually prepare a little more than I need for dinner.

Grilled Pork Loin Roast

2 T pepper
1 boneless pork loin roast (4-5 pounds)
$1/2$ C packed brown sugar
1 T cornstarch
$1/4$ t ground cinnamon
$1/4$ t ground cloves
$1/8$ t ground ginger
$1/2$ C pineapple juice
$1/2$ C sweet and sour sauce
$1/4$ C Worcestershire sauce
3 T lemon juice
apple and orange wedges.

Rub pepper over top of roast. Grill, covered, over low heat for $2^1/2$ to 3 hours. Let stand 10 minutes before slicing.

In a saucepan, combine the brown sugar, cornstarch and spices.

Stir in the pineapple juice, sweet-and-sour sauce, Worcestershire sauce and lemon juice until smooth.

Bring to a boil, cook and stir for 2 minutes or until thickened.

Garnish the roast with apple and orange wedges and serve with the sauce.

POTS, PANS, PONIES & PINES

Beef Stew

3 lbs. beef stew meat
6 carrots, peeled and cut into 1 inch pieces
6 small potatoes, peeled and cut into 1 inch pieces
1 pkg. pearl onions, (8 oz.) frozen
1 C red wine
2 C boiling water
flour, salt and pepper

Dust stew meat with flour, salt and pepper.
Brown in hot oil or lard in medium Dutch oven.
Add 1 C red wine, and 2 C boiling water.
Place Dutch oven in hot pit (coals under oven).
Bank coals on sides and on top. Simmer for 2 hours.
Add carrots, potatoes and thawed pearl onions and simmer for another hour or until vegetables are tender.
Remove oven and place biscuits on top (either make from scratch or use Pillsbury Hungry Jack), put the lid back on the Dutch oven, place coals on top and cook for an additional 12-15 minutes.

This has to be buried in coals to cook the vegetables. If you are not able to use a Dutch oven, partially cook the carrots and potatoes before adding to the stew.

Cast Iron Cooking on the Trail

Veal Chops*

4 veal chops
$^1/_2$ flour
$^1/_2$ C butter
1 bunch green onions
2 C mushrooms.
$^1/_2$ C white wine

Flour chops only on one side.

Brown the chops on both sides in 3 T melted butter in large skillet using the grate.

Shovel hot coals under the grate for heat source.

Adjust as necessary.

Remove chops and set aside.

Sauté green onions and mushrooms in remaining butter and $^1/_2$ C white wine.

Put chops back in skillet, partially cover and cook approximately 8 minutes on each side, turning 2-3 times and basting often.

Mexican Casserole*

6 to 8 flour tortillas
1 large can (1 lb. 12 oz.) enchilada sauce
(either red or green depending on taste)
2 lbs. hamburger
1 onion, chopped
2 C grated or shredded sharp cheese

Brown meat.
Heat sauce.
Dip tortillas in sauce and put the first layer of tortillas in bottom of large skillet.
Layer with beef, chopped onions and cheese.
Repeat with another layer of tortillas and other ingredients.
Pour what is left of everything including sauce over top.
Cover the skillet with foil, top with foiled Jackson Grill and place in cool pit (no coals under the bottom of the skillet).
Bank sides and top of Jackson Grill with medium coals. Check after 20 minutes. Cook until warm and bubbly.

Cast Iron Cooking on the Trail

Lasagna

1 pkg. lasagna noodles
2 lbs. Italian sausage   1 lb. lean hamburger
1 large can (1 lb. 12 oz.) diced Italian tomatoes
1 can (12 oz.) tomato paste
(Use same can and add an additional $1/2$ can water)
3 C Ricotta cheese (1 lb. 14 oz.)   4 C grated mozzarella cheese
3 eggs, beaten   Parmesan cheese

Brown meat and drain grease.

Add tomatoes and paste with water.

Simmer for 1 hour or until most of liquid is gone.

Cook lasagna noodles. Do not overcook, they will cook in the Dutch oven as well.

Mix the ricotta cheese with beaten eggs.

Line a large Dutch oven with heavy duty foil and layer ingredients as follows:

Starting with the noodles, then meat sauce, ricotta cheese and egg mixture, mozzarella cheese and Parmesan cheese. Repeat until all ingredients are gone.

Cover with foil. Put on lid. Bury the Dutch oven in cool pit (no coals on the bottom) bank medium coals around side and on top. Cook for 1 hour, but check after 30 minutes. Adjust heat if necessary and rotate the oven.

This can be assembled and frozen in the Dutch oven at home. I use a Dutch oven that has a flat bottom, not one that has the legs. I have the wranglers top pack it. They tie it on top of the mule or horse with lots of padding. It is a great and easy first night meal. When you get to camp, build a fire, get your coals ready and bury the Dutch oven. It saves a lot of time if you have a long riding day.

Grandma's Meat Pie

1 lb. ground sirloin (or very lean ground beef)
salt and pepper to taste

Pie Crust:

2 C flour
1 C lard or Crisco
4 T cold water
pinch of salt.

Put water in with ground beef, just to cover.
Simmer for 45 minutes in a saucepan.
Season to taste. Add one T Worchester sauce.
Add 1 T flour to thicken.
Take $1/2$ of raw piecrust and roll out. Put in pie pan.
Pour in meat filling.
Roll out top crust and cover the pan.
Make several slits in the crust.
Put an inverted pie pan in Dutch oven, place meat pie on inverted pie pan. Put the lid on.
Place oven warm pit (a few coals under the Dutch oven) and bank the sides and top with hot coals. Check often and cook until the crust is brown.
The right temperature can be tricky with pies. Check the temperature often. Do not put too many coals beneath the Dutch oven or around the side because the crust will burn.

Cast Iron Cooking on the Trail

Meat Loaf

1^1/2 lb. lean ground beef
1 C breadcrumbs or 1 C ground saltine crackers
2 beaten eggs
1 small onion-minced
1 jar (4.5 oz.) mushrooms
^{1}/2 C salsa   salt and pepper to taste

Mix all ingredients. Pour salsa on top of meat. Put in small Dutch oven and bury in pit with medium coals. Place a few under the oven, bank the sides and place a few on the top of the lid.

Check after 20 minutes. Continue to cook until meat is light pink. Remember the meat will continue to cook after you remove it from the heat. So if you like it medium rare, take it out a little early.

Easy Pot Roast

4 pounds chuck roast
1 envelope dry onion soup mix
1 can (10.5 oz.) condensed cream of mushroom soup
6 potatoes and 6 carrots, cut in 2 inch cubes

Place a large piece of heavy duty foil in Dutch oven. You may need to fold two pieces of foil together to get the length you need.

Place meat on foil. Sprinkle soup mix over top of meat and spread with cream of mushroom soup.Add potatoes and carrots.Fold over the foil and secure tightly.

Place foil over the top and add lid. Bury Dutch oven in medium coals. Place a few coals underneath, bank the sides and the top. Check often and rotate the oven.

Shoud cook for 2 hours.

Pot Roast with Vegetables

2 C water
2 C vinegar
4 medium onions, quartered
12 whole peppercorns
4 bay leaves
4 whole cloves
2 T salt
1 T Worcestershire sauce
1 t pepper
$1/2$ t garlic powder
1 boneless beef, venison, or elk rump or chuck roast ($3^1/2$ to 4 lbs.)
2 T vegetable oil
10 medium carrots cut into 1-inch chunks
5-7 T cornstarch
$1/3$ C cold water

In large plastic bag or large bowl, combine the first 10 ingredients; mix well.

Add the roast. Cover and chill for 24 hours.

Put this in a pannier that you are hanging. Let it chill overnight.

Take out the roast, reserving the liquid.

In a Dutch oven, brown the roast in oil, then drain excess oil.

Add the marinade and the carrots; bring to boil.

Bury Dutch oven in pit with medium coals on bottom, sides and top for $3^1/2$-4 hours until the meat is tender.

Check often to check on temperature of coals.

Remove roast and keep warm. Remove vegetables and spices.

Combine cornstarch and cold water until smooth then gradually add to pan juices for gravy.

Bring to a boil and stir for 2 minutes.

Slice roast and serve with gravy.

Beef Brisket

Sauce:

$2^1/4$ C ketchup
$1^1/2$ C beer
$1/2$ C packed brown sugar
$1/2$ C wine vinegar
2 T Worcestershire sauce
2 T chili powder
3 garlic cloves, minced
1 C chopped onion
$1/4$ t cayenne pepper
1 (5 to 7 pound) beef brisket with fat trimmed
2 T liquid smoke

Combine ketchup, beer, sugar, vinegar, Worcestershire sauce, chili powder, garlic, onion and cayenne pepper and bring to a boil.

Reduce heat and simmer for 30 minutes or until sauce thickens.

Pour 2 C of sauce into bowl and let cool.

Brush both sides of brisket with liquid smoke and place in large Dutch oven.

Pour remaining sauce over brisket, turning to coat evenly. Cover with foil and put lid on.

Bury Dutch oven; placing a few coals underneath, bank sides and put on top of lid. Use medium coals.

Check often to maintain temperature.

Cook for 4 hours.

Remove brisket from pan juices, slice thinly across the grain. Heat reserved sauce and serve over sliced brisket.

Tom's Shortribs:

3-4 lbs. short ribs
2 T oil
¼ C flour
salt and pepper to taste

Dust ribs with flour, season with pepper and salt.
Brown in small amount of oil in medium Dutch oven.

Add:

1 can (10.5 oz.) tomato soup
1 C red wine
1 C consommé or bouillon
3 peeled quartered potatoes
2 quartered onions

Cover with foil, add lid.
Bury Dutch oven in warm pit.
Place a few medium coals under the bottom, bank sides and add coals to the top of lid.
Cook for four hours and check often to maintain temperature.
Rotate Dutch oven every hour.

Cast Iron Cooking on the Trail

Brown Sugar and Bourbon Ribs

2 , 2$\frac{1}{2}$ lb. racks baby back ribs (Total: 5 lbs.)

Rub:

1 T coarse kosher salt
1 T brown sugar
1$\frac{1}{2}$ t dry mustard
1$\frac{1}{2}$ t dried thyme
1 t ground ginger
$\frac{1}{2}$ t ground cinnamon
$\frac{1}{2}$ t cayenne pepper

Roasting sauce:

1 large onion, sliced
1 t cinnamon
1 t ground ginger
$\frac{1}{4}$ C vinegar

Basting sauce:

$\frac{1}{2}$ C brown sugar
2 T Dijon mustard
$\frac{1}{2}$ C butter
$\frac{1}{4}$ C bourbon whiskey
$\frac{1}{4}$ C apple cider vinegar

At home, mix the first seven ingredients and rub into each side of ribs. Freeze at home and then let them thaw in camp. You want them to have chilled for several hours.

Place onion, 1 t cinnamon and 1 t ginger in large skillet.

Pour in vinegar.

Add ribs, meat side down, and cover skillet with foil and foiled Jackson Grill.

Put skillet in warm fire pit with a few coals underneath the skillet, around sides and on top of Jackson Grill. Roast on low to medium heat for 2 hours.

Uncover and cool.

In the meantime, combine ingredients in basting sauce and bring to a boil.

Grill ribs on grate using coals underneath.

Adjust heat as needed. Grill until heated through and slightly charred (about 5 minutes on each side.)

Brush generously on all sides with basting sauce.

Cast Iron Cooking on the Trail

Grilled Sirloin Steak

1 boneless beef sirloin steak (3 lbs. and 2 inches thick)
2 onions, sliced
1 lb. fresh mushrooms, sliced

Marinade:

1 C soy sauce
$1/4$ C red wine vinegar
$1/4$ C olive or vegetable oil
4 chopped garlic cloves
1 T pepper
1 T ground ginger
1 T honey

Sauté mushrooms and onions in 2 T butter (each) in separate pans

Combine marinade ingredients in pan. Add steak and turn once. Marinate for at least 3 hours. Drain and discard marinade.

Grill steak on grate over medium to hot coals for 10 minutes on each side. Remove and cover with foil if you want it more well done. Let rest for 2 to 3 minutes.
Slice and serve with onions and mushrooms on the side.

Stuffed Tenderloin

3 lbs. beef tenderloin.
2 T oil or butter

Stuffing:

6 T butter
$^1/_2$ C chopped onion
$^1/_2$ C chopped celery
3 C bread cubes
1 small jar (4.5 oz.) sliced mushrooms, drained
3 uncooked bacon slices

Slice tenderloin in half (lengthwise).
Place in skillet. In 2 T oil, brown each side, top and bottom.
Remove from skillet.
Sauté onion and celery in the butter.
Stir in bread cubes and add mushrooms.

Place meat in Dutch oven. Spread stuffing over half the meat.
Fold top over. Season meat with salt and pepper.
Top with 3 uncooked bacon slices.
Cover Dutch oven with foil, put lid on top and bury with medium coals in warm pit. Place a few coals under the oven, then bank coals around side and place a few on top.

Cook for 1 hour, checking and rotating oven every 20 minutes.
Slice and serve with large salad and crusty bread.

Cast Iron Cooking on the Trail

Prime Rib

1 4 to 6 lb. standing rib roast
salt, pepper

Season prime rib with salt and pepper.

Place in Dutch oven, cover top of Dutch with foil and put on the lid. Bury in warm pit with medium coals. Put a few coals underneath, bank sides and top.

Cook for 40 minutes, checking temperature after 20. Remove from coals when meat appears rare in the middle. Meat will continue to cook, so it is better to remove the meat from the Dutch oven and then tent it with a piece of foil. Let rest for 2-4 minutes and then check again.

Liver and Onions

2 lbs. liver
1 lb. pkg. thick sliced bacon.
1 large onion, sliced

Fry bacon and sliced onions in bacon grease. Set aside.

Lay slices of liver in greased skillet.

Cover with onions and lay bacon on top.

Add a small amount of water.

Simmer 15 minutes, adding small amounts of water as needed to avoid drying.

Joe's Elk Tenderloin

3 to 4 lbs. elk tenderloin
(usually sliced $1/2$ inch thick in 6 or 8 pieces)
2 T butter
$1/2$ C Heinz 57 sauce
2 T Worchestershire sauce
2 T yellow French's mustard
$1/2$ C red wine
$1^1/2$ sliced medium onions
1 can (4.5 oz.) mushrooms
garlic
salt and pepper

Brown onions in butter. Set aside.
Add remaining ingredients to pan, except for meat.
Simmer this sauce until bubbly.
Add lightly peppered elk slices.
Cook slowly for about 20 minutes.
Add onions and mushrooms, simmer for additional 5 minutes.

Cast Iron Cooking on the Trail

Shepherd's Pie

2 lbs. ground beef or ground lamb
1 large onion, finely chopped
4 carrots, coarsely chopped
2 T tomato paste
2 T flour
1 to 2 T Worcestershire sauce
Salt and pepper to taste
10 oz. frozen peas, thawed
$2^1/2$ lbs. russet potatoes peeled and quartered
1 C milk
6 T butter

Brown meat and drain well.
Add ¼ C water to the skillet, scraping up browned bits.
Add onion and carrots and cook.
Stir occasionally, until softened, about 5 minutes.
Stir in tomato paste. Add flour, cook, stirring for 2 minutes.
Add Worcestershire sauce, 2 C water and meat.
Season with salt and pepper.
Simmer until thickened, about 10 minutes.
Stir in peas. Remove from heat.

Meanwhile, cover potatoes with water in medium saucepan and bring to a boil.
Reduce heat and simmer until fork tender, 15 to 20 minutes. Drain.
In a pan bring milk and butter to a simmer, add potatoes and mash. Season with salt and pepper.
Spread over top of meat mixture. Cover with foil, top with foiled Jackson Grill. Top with coals. Bake for 25 to 30 minutes until potatoes are brown and dish is hot.

Chicken Fried Steak*

One cubed steak for each person

Flour steak, then dip in egg and milk, and finish with breadcrumbs. Repeat (double dip!)
Melt 2 T butter in large skillet and fry cubed steak 5 to7 minutes on each side in hot skillet
Serve with mashed potatoes and gravy

Ham

4-6 lb. precooked ham

Glaze:

$^1/_2$ C butter
$^1/_2$ C brown sugar
$^1/_2$ C bourbon
small jar (6 oz.) currant jelly

Place pre-cooked ham in foil-lined Dutch oven
Melt the butter, add the brown sugar and bourbon and pour over ham, cover with foil and add lid.
Place in warm pit. Put a few medium coals underneath, bank sides and place a few coals on top. Since the ham has been pre-cooked you only need medium to low coals for 30 to 40 minutes.

Cast Iron Cooking on the Trail

Meat and Potatoes Kabobs

1$^1/_2$ lb. sirloin steak, cut in 2 inch cubes
3 T soy sauce
2 garlic cloves, smashed and chopped
16 small new round potatoes, peeled and parboiled
2 T catsup
1 T honey
salt and pepper to taste

Place beef, soy sauce, garlic and oil in bowl.
Add meat and marinate 15 minutes.
Alternate beef and potatoes on skewers.
Mix catsup with honey and brush over skewers
Season generously.
Cook on grate over hot coals 7 minutes on each side.
Adjust the grate and coals as necessary for proper heat.
Rotate skewers several times while cooking.

POTS, PANS, PONIES & PINES

Strips of Beef and Vegetables

2 T olive oil
$1^1/_2$ lb. beef sirloin tip
3 garlic cloves, smashed and chopped
1 head broccoli (in florets), blanched 4 minutes
16 large cherry tomatoes
$^1/_2$ red onion, cut in 2 and sectioned
juice of 1 lemon
salt and pepper

Heat 1 T oil in skillet. When very hot, add the whole piece of meat and sear on all sides. Season well.

Slice beef into $^1/_2$ inch strips.

Place in bowl along with remaining oil, garlic and lemon juice. Marinate 15 minutes.

Drain beef and reserve marinade.

Fold the pieces of meat in half and alternate on skewers along with vegetables.

Broil on the grate over hot coals.

Cook six minutes on each side.

Baste several times with marinade.

Cast Iron Cooking on the Trail

Pineapple Chicken Kebobs

2 chicken breasts, skinned, boned
1 large can (1 lb. 12 oz.) pineapple chunks
5 slices cooked bacon, cut in half
3 T butter
2 garlic cloves, smashed and chopped
2 T chopped parsley
A few drops of Worcestershire sauce
salt and pepper to taste

Cut chicken in 2 inch pieces.
Alternate along with pineapple and bacon on skewers.
Melt butter in small pan, over medium heat.
Stir in garlic, parsley and Worcestershire sauce.
Baste skewers with melted butter mixture.
Season with salt and pepper

Cook over grate over medium coals for about 12 minutes.
Adjust your grate and number of coals to adjust your heat.
Turn skewers once or twice.

Cast Iron Cooking on the Trail

Vegetables & Side Dishes

Cast Iron Cooking on the Trail

POTS, PANS, PONIES & PINES

Have you noticed there are a great variety of frozen vegetables now available in grocery stores?

Take advantage of the convenience and quality of produce in the frozen section. Another plus is the frozen packages will also keep your cooler panniers cold.

If you are not using the vegetables in a specific recipe, you can "dress them up" by adding butter, red pepper flakes, sliced almonds, sliced mushrooms and all types of seasoning. Experiment with the flavors you like.

If you don't want to pack in a lot of butter, try Butter Buds or a similar product. Just add water and you have a good butter-flavored substitute.

Cast Iron Cooking on the Trail

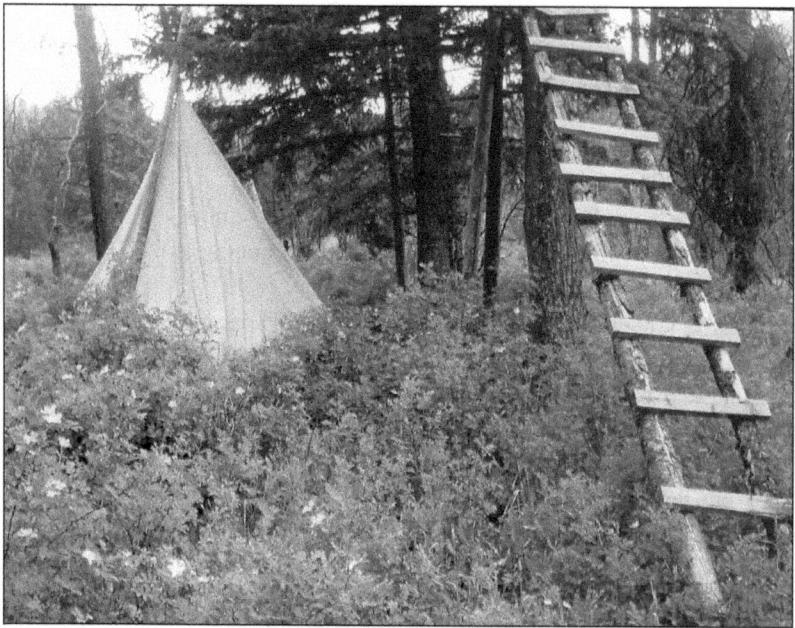

Elks Fork. The ladder goes up to a food cache
and is taken down at night to prevent bear trouble..

Garlic Grits

$4^1/2$ C boiling water
1 C instant grits
1 t salt
1 roll (6 oz.) garlic cheese (found in the Deli section)
1 stick butter
2 eggs

Boil water. Add grits and salt. Cook according to directions on grit package, but usually 12-14 minutes until thickened.

Add garlic cheese broken into small pieces (or use $^1/2$ C Velveeta cut into small cubes and seasoned with garlic powder to taste)

Add butter after the grits are cooked. Beat 2 eggs lightly in a one cup measure, add milk to fill cup and add to grits.

Pour mixture into Dutch oven. Place in pit with medium coals (no coals underneath). Bank coals around side and on top. Bake 30 minutes until grits have set and top is lightly browned.

Cast Iron Cooking on the Trail

Mashed Potatoes

2 pounds potatoes
$1/2$ C milk
$1/2$ stick butter

Peel, wash and quarter potatoes. Place potatoes in large pan, cover with cold water. Place pan on grate. Bring to boil, reduce heat to medium and simmer until tender, about 20 minutes. Drain well. Mash potatoes. Stir in butter. Warm the milk and add.

Variation: Prepare mashed potatoes but do not add the milk. Stir together 1 C sour cream, $1/3$ C chopped chives and $1/4$ C prepared horseradish in medium bowl. Add to potatoes and season with salt.

Roasted Red Potatoes

1 pkg. (1 oz.) Hidden Valley original ranch dressing mix
3 pounds small red potatoes, par boiled and cut in half
$1/4$ C vegetable oil

Place potatoes in large plastic bag. Add oil and toss to coat. Add dressing mix and toss again. Use a vegetable grill to roast on grate and turn often or fry in skillet until tender.

Variations:

Place in bag with 2 T olive oil, parsley, fresh oregano, garlic, green onions and Parmesan

Sprinkle with $1/2$ C sea salt and $1/2$ C vinegar

POTS, PANS, PONIES & PINES

Baked Potato and Onion Pie

4 lbs. white potatoes, peeled and thinly sliced
2 large onions, thinly sliced
$^1/_2$ C butter
$^1/_2$ C grated Parmesan cheese

Layer ingredients in large skillet. Cover with foil and foiled Jackson Grill.

Put skillet in warm pit. Put a few coals underneath, on top and around the sides of the skillet. Cook until tender, about 35 minutes but check after 15 minutes and rotate skillet.

Bacon, Onion and Cheese Potatoes

8 slices bacon, coarsely chopped
2 lbs. russet potatoes, peeled, cut into $^1/_2$ inch pieces
1 large onion, chopped
1 C packed grated pepper jack cheese
2 T butter

Place medium skillet on grate.

Cook chopped bacon in skillet over medium heat until crisp and brown.

Transfer bacon to paper towels and drain. Return bacon, along with potatoes and onions to skillet. Season with salt and pepper. Sprinkle cheese on top and dot with butter.

Cover with foil and foiled Jackson Grill lid. Place skillet in warm pit. Put a few coals underneath the skillet, around sides and add a few on top of the Jackson Grill.

Bake 40 minutes or until potatoes are tender. Check every 15 minutes and rotate skillet. Adjust heat as necessary.

Cast Iron Cooking on the Trail

Cheesy Potatoes

2 lbs. frozen cubed hash browns, partially thawed
1 C finely diced onions
1 can (10.5 oz.) cream of celery soup
$1/4$ C melted margarine
1 large container (24 oz.) sour cream
8 oz. grated cheddar cheese

Topping:

2 C crushed cornflakes
$1/2$ C melted butter (1 stick)

Mix onions, soup, margarine, sour cream, cheddar cheese and salt and pepper. Add to partially thawed hash browns.

Line skillet with foil. Put potato mixture in skillet.

Mix corn flakes with melted butter. Spread on top of potatoes and cover with foil and foiled Jackson Grill. Place skillet in warm pit, putting a few coals underneath the skillet, bank the sides and add coals to the top of the Jackson Grill.

Cook 50 minutes but check at 30. Rotate skillet and adjust heat. Top should be crusty brown and potatoes should be tender.

Potato Wedges

$^1/_3$ C all purpose white flour
$^1/_3$ C Parmesan cheese
1 t paprika
3 large baking potatoes, par boiled an cut in 8 wedges each
$^1/_3$ C milk
$^1/_2$ C butter

Sour Cream Dip:

2 C sour cream
8 bacon strips, cooked and crumbled
2 T chives
$^1/_2$ t garlic

In plastic bag, combine first 3 ingredients. Dip potato wedges in milk. Place in bag and coat. Place in medium skillet and put skillet on the grate.

Drizzle with half of the butter and fry uncovered for 20 minutes on high heat, turning often. Drizzle with remaining butter and continue to cook until potatoes are tender and have a crusty brown exterior. Serve with dip.

Cast Iron Cooking on the Trail

Potatoes Au Gratin

6 medium potatoes peeled and parboiled
$1/2$ C chopped onion
2 large cloves garlic, minced
2 T olive oil
$1/4$ C flour
1 t salt
$1/4$ t pepper
3 C milk
1 C shredded Parmesan cheese

Cook potatoes in slightly salted boiling water for 15-20 minutes. Drain and slice thin.

In saucepan cook onion and garlic in hot oil over medium heat until tender. Stir in flour, salt and pepper. Add milk all at once. Cook and stir until bubbly.

In medium Dutch oven or skillet, layer half the potatoes, sauce and cheese. Repeat with remaining potatoes and sauce.

Put Dutch oven in medium coals, spreading a few coals underneath the oven, around the sides and few on top.

Cook until potatoes are done and the top is a crusty brown, 35 minutes but check after 15 and rotate the oven.

Onion Casserole

$1/4$ C butter
7-8 large onions, sliced
$1/2$ C instant rice, cooked
1 C Swiss cheese
$1/2$ C half and half
Salt

Place onions and butter in large skillet on the grate. Cook until onions are translucent. Mix in remaining ingredients. Cover with foil and cook until the cheese is melted.
Serve with any type of meat

Ginger Carrots

8 large carrots, peeled and cut in $1^1/2$ inch pieces
$1/2$ t salt
$1/2$ C honey
4 T sugar
4 T margarine
1 t grated lemon rind
$1/4$ t ground ginger

Put carrots in medium saucepan and place on grate. Add water and salt to cover and put lid on top. Cook covered for 10 minutes. Drain and add honey, sugar and margarine. Cook uncovered, stirring occasionally, until carrots are tender. Stir in lemon rind.

Cast Iron Cooking on the Trail

Dutch Oven Cabbage

1 head cabbage
1 lb. bacon
1 onion, chopped
1 C water
1 T Chili powder
or 1 can (4 oz.) diced green chilies (hot fire roasted)
salt and pepper to taste

Slice bacon in 1 inch pieces and fry in a hot Dutch oven.
When partially done, add one chopped onion.
Cook the onion until tender and add water.
Slice the cabbage in thin wedges and add to the pot. Cover.
Keep at a slow boil until done. (10-15 minutes)
Add chili powder or green chilies. Salt and pepper to taste.

Serve with ham.

Spanish Rice

4 slices bacon, crumbled
1 medium green bell pepper, diced
1 small onion, chopped
1 large garlic clove, diced
$1/2$ lb. ground beef
1 can (14.5 oz.) stewed tomatoes
1 C long grain rice
1 C water
$1^1/2$ t Tabasco
1 t salt

In medium skillet, on grate over medium heat, cook bacon until crisp, stirring occasionally. Drain and move to paper towels.

In drippings, cook green pepper, onion and garlic until tender and crisp (about 5 minutes). Remove to bowl. Set aside.

In same skillet, over medium heat, cook ground beef until well browned on all sides, stirring frequently.

Add tomatoes with their liquid, rice, water, Tabasco sauce, salt and green pepper mixture.

Over high heat, bring to boil, reduce heat to low.

Cover and simmer 20 minutes or until rice is tender, stirring occasionally.

Cast Iron Cooking on the Trail

Easy Baked Beans

6 slices bacon
2 large cans (1 lb. 12 oz.) pork and beans
1 t dry mustard
2 small onions, finely chopped
$1/2$ C ketchup
$1/2$ C molasses
salt and pepper to taste

Fry bacon in Dutch oven until partially cooked, almost done but not crisp. Cut each slice into fourths. Add remaining ingredients. Mix. Bake in warm pit (place a few coals underneath Dutch oven), bank coals on sides and on top.

Cook on very low heat for 2-3 hours. Check often and rotate Dutch oven.

Broccoli and Rice

1 pkg. frozen broccoli (10 oz.) thawed
$1^1/2$ C Velveeta cheese or small jar (10 oz.) Cheese Whiz
1 small onion, diced
1 C cooked instant rice
1 can (10.5) mushroom soup
$1/2$ C milk

Place broccoli in medium skillet after draining excess water.

Add cheese, onion, rice and mushroom soup. Mix in skillet and place on grate. Cover with foil.

Cook over medium heat until soup is bubbling and cheese is melted. Add milk and stir.

Confetti Rice

1 C chopped green onions
1 medium size green pepper cut into $1/2$ in. cubes
1 medium size red bell pepper, cut into $1/2$ in. cubes
2 minced garlic cloves
$1/2$ t oregano
$1/4$ t salt
1 t cumin
2 T butter
3/4 C uncooked white rice (not instant)
$1/4$ C water
1 can (10.5 oz.) chicken broth

Place in medium skillet on grate. Sauté first 8 ingredients over medium heat for 3 minutes.

Add water and chicken broth.

Bring mixture to a boil.

Add rice. Cover, reduce heat and simmer 20 minutes or until rice is tender and liquid is absorbed.

Cast Iron Cooking on the Trail

Green Chilies and Rice

1^1/$_2$ C rice
^{1}/$_3$ C chopped green onions, including tops
1 T butter
1 C shredded cheddar cheese
1 can (4 oz.) diced green chilies, undrained
1 green pepper cut into matchsticks
1 medium chopped tomato
1 medium red onion, diced
5 eggs, beaten
^{1}/$_3$ C milk
1 t Worcestershire sauce
^{1}/$_2$ t salt
3 to 4 drops Tabasco

In medium skillet over medium heat, sauté onions in butter until tender but not brown.

Combine rice, cheese, chilies, bell pepper and tomato in large bowl. Add cooked onions.

Combine eggs, milk Worcestershire sauce, salt and pepper and Tabasco in small bowl.

Pour into skillet. Stir in rice mixture.

Cover with foil and top with foiled Jackson Grill. Place skillet in warm pit, (no coals underneath skillet), bank coals on sides and on top of Jackson Grill.

Bake in medium heat for 25 to 30minutes or until rice is done. Check at 15 minutes.

Rotate skillet and adjust heat as necessary.

Vermicelli Noodles

1 pkg. (12 oz.) vermicelli noodles
1 carton (24 oz.) sour cream
2 T chopped dried chives
1 small container (12 oz.) cottage cheese
$1/2$ C butter (1 stick)
salt and pepper to taste

Cook vermicelli noodles according to package directions. Add sour cream, chives and cottage cheese. Salt and pepper to taste.

Line medium skillet with foil. Put noodle mixture in skillet and dot generously with butter.

Cover with foil and foiled Jackson Grill. Place skillet in fire pit. Place coals around sides and put coals on top of Jackson Grill.

Bake for 30 minutes, but check at 15 and rotate skillet.

Fettuccine Alfredo

1 lb. fettuccine, cooked al dente according to package directions
$1/2$ C butter
1 egg yolk
$1/4$ C cream
$1/2$ C grated Parmesan cheese
Salt and pepper to taste

Boil fettuccine in lightly salted boiling water. Drain, but leave a little bit of pasta water. Cover and set aside. Cream butter, add egg yolk and cream and beat while adding cheese, one T at a time. Add mixture to hot Fettucine. Serve immediately

Old Fashioned Bread Stuffing

1^1/2 C chopped or sliced celery
1 C chopped onion
^{1}/2 C butter
1 T snipped fresh sage or 1 t poultry seasoning
^{1}/4 t black pepper
12 C dry bread cubes (use store bought)
1 to 1^1/4 C chicken broth

In a large skillet cook celery and onion in hot butter over medium heat until tender. Remove from heat. Stir in sage and pepper.
Place bread cubes in large bowl and add onion mixture.
Drizzle with enough chicken broth to moisten, and toss lightly to combine.
Place stuffing in small Dutch oven and put in warm pit. Place a few coals underneath, bank coals around sides and place more on top.
Bake covered for 25-30 minutes or until heated through. Check after 15, fluff the dressing and rotate the Dutch oven.

Variations:

Stir 2 medium cored apples chopped into bread cubes.

Stir one 15 oz. can chestnuts, drained and coarsely chopped into the bread cubes.

Stir 1 C cooked wild rice into the bread cubes.

Omit 1 C celery and substitute 2 C sliced mushrooms, cooked. Add to bread cubes.

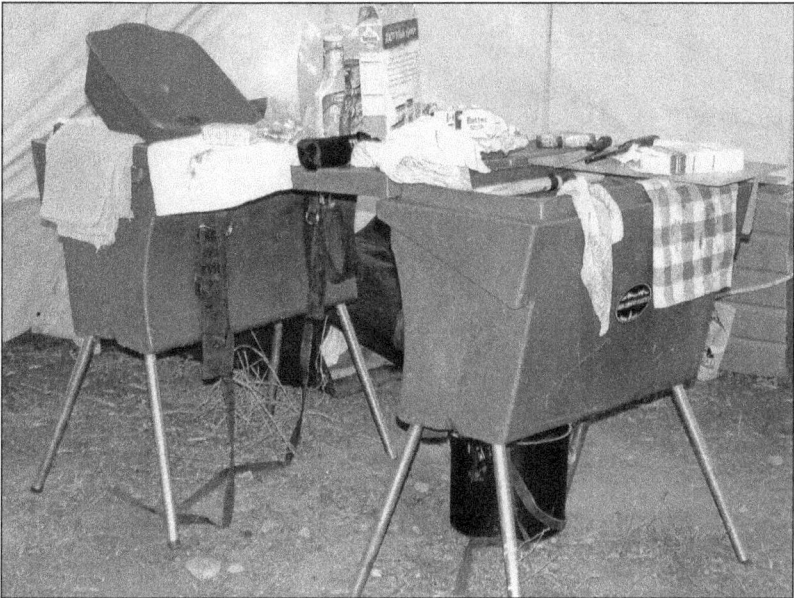

Kitchen paniers that connect with a shelf for meal preparation

Desserts

Lemon cake with raspberries..

POTS, PANS, PONIES & PINES

Desserts are fun to make in the mountains, but be sure to allow sufficient time.

I try to bake after breakfast. By the time breakfast dishes have been done, the breakfast pit has cooled down. Remember you have been burning wood and making coals to cook breakfast meat, pancakes, coffee, etc. This makes for a pretty hot pit. Keep a small feeder fire going so you can add coals to bake your desserts, but you won't need a huge fire as the pit is already warm.

If your pit is totally cold because you are baking in the afternoon, place a few hot coals where you are going to put your Dutch oven or skillet. Let these cool just a little before putting your pan in the pit. Remember to use your hand to gauge the temperature of the coals. If you are 3-4 inches away and it is really hot, it is similar to 400-450 degrees in a conventional oven. If it is comfortable, but hot, it is about 350 degrees and if it is warm, it is low heat, 250 degrees for a conventional oven.

Also, when you are checking the temperature and rotating the skillet or cake pan, be careful to let as little heat as possible out of the pan or Dutch oven. Think of a cake falling when you open up your oven door suddenly.

Smell what is cooking. If it smells like it is burning, it probably is. Check it immediately and adjust your coals.

Another thing to remember when using coals as the heat source is to wrap the Jackson Grill in foil when you are using it for a skillet lid. This will keep it clean when you use it for a regular grill.

Cast Iron Cooking on the Trail

Baking in hunting camp with a wood burning stove is similar to being at home. Pack in a thermometer to put in the oven and you will be fine. Or if you are really lucky you might have an oven pannier on summer trips. Again, an oven thermometer really makes your baking pretty easy. Metal baking pans and cookie sheets work great in wood burning stoves or propane ovens. But if you are baking with coals, use cast iron skillets and cast iron Dutch ovens.

If a recipe says to chill a dessert, put it on the inside of a cooler pannier or make a rock barrier on the edge of a stream if possible. The shallow stream water flowing under the pan will keep it cold.

I can't emphasize this enough: Have patience and have fun. Check your batter or crust every fifteen minutes or so and rotate your pan when you add coals. Your cakes will get done in due time. And remember it's fun to experiment. Don't be upset if the desserts don't turn out perfectly. Remember you're cooking in the mountains with coals. Make a note of what goes wrong and change it the next go around.

Apple Pie

Crust for a 2 crust pie:
2 C flour
1 C Crisco or lard
4 T cold water
pinch of salt

Filling:

6 medium Granny Smith apples, partially peeled and sliced
2 T flour
$^1/_4$ C sugar
1 t nutmeg
1 t ground cinnamon

Mix crust ingredients with a fork until dough looks like small peas. Divide this mixture and form into 2 balls. Set aside.

For the filling mix flour, sugar, nutmeg and cinnamon and sprinkle over the apples. Set aside.

Roll out one ball uncooked crust on flat, floured surface to the size of a dinner plate. (I have a 6 inch rolling pin that I take in with my cooking gear.) Put the crust in a small cast iron skillet.

Put apple mixture in crust.

Slice 3 T butter and put on top of apples.

Roll out second ball of crust and put on top and pinch around edge of skillet to seal both crusts. Push the top of the crust down into the skillet, so the pie looks more like a deep dish pie.

Cover with foil and top with Jackson Grill. Put coals on top of foiled Jackson Grill and around sides of skillet. Bake until apples are done and crust is brown, about 30 minutes.

Check after 10 minutes and rotate the skillet often.

Cast Iron Cooking on the Trail

Variations:

Fill with any type of available fruit. If using frozen fruit, thaw, simmer a few minutes and add 1 T cornstarch until mixture thickens.

Place a few medium coals on bottom if you are starting with a cold pit. Spread hot coals around edge and on top. Remember to check after 10 minutes. Coals have to be hot to cook crust and fruit, but as the skillet heats up, you might have to reduce the number of coals.

Apple Crisp

8 medium Granny Smith apples (usually one per person).
1 T cinnamon
1 cube butter
$^1/_2$ C brown sugar
$^1/_2$ C flour
4 to 6 small pkgs. instant oatmeal with cinnamon
$^1/_2$ C water

Partially peel, core and slice apples in thin pieces.

Peel in spiral pattern which will leave part of the peel on. It gives the crisp a nice flavor.

Mix cinnamon and water with apples and place in bottom of the skillet.

Mix butter with sugar, flour and oatmeal until crumbly. Place on top of apples.

Cover with aluminum foil. Top with foiled Jackson Grill.

Place in warm pit, covering top and sides with coals.

Bake until apples are done (about 45 minutes).

Serve with whipped cream.

I buy the 13 oz. whipped cream in a jar.

Applesauce Brownies

$1/2$ C shortening
2 squares unsweetened chocolate
1 C sugar
2 well beaten eggs
$1/2$ C applesauce
1 t vanilla
$1/2$ C chopped nuts

Sift together:
1 C flour
$1/2$ t baking powder
$1/4$ t salt
$1/4$ t soda

Melt shortening and chocolate in medium saucepan. Take off heat and set aside.

Blend sugar, eggs, applesauce and vanilla. Add this mixture to chocolate and stir.

Add the flour mixture and stir. Add the nuts.

Line cast iron skillet with aluminum foil. Pour in batter. Cover with foil. Top with foiled Jackson Grill.

Place in warm pit, (no coals under bottom of skillet), bank sides with medium coals and put coals on top of Jackson Grill. Bake until firm 35 to 40 minutes. Check at 20 minutes and then rotate.

Spoon Hershey's chocolate syrup and Cool Whip over brownies.

Applesauce Cake

$2^1/_2$ C applesauce
4 t baking soda
2 C raisins
2 C brown sugar
1 C shortening
2 eggs
3 C flour
1 t cinnamon
1 t nutmeg
$^1/_2$ t salt
1 C nuts, chopped

Cream shortening and brown sugar.
Add eggs and mix well.
Stir raisins, applesauce, and soda together.
Sift dry ingredients and add to creamed mixture.
Stir in nuts. Place in 9 x 13 pan or medium skillet.
Put foil over the skillet and top with foiled Jackson Grill.
Cook in warm pit, (no coals under the pan) bank sides with medium coals and place a few on top of the Jackson Grill.
Bank until cake is firm, about 50 minutes, but check and rotate at 30 minutes.

Cast Iron Cooking on the Trail

Peach Oat Crumble

1 large can (1 lb. 13 oz.) sliced peaches, drained
2 T lemon juice
$1/4$ t cinnamon
1 T butter
$1/4$ C solid vegetable shortening, melted

Topping:

$1/3$ C brown sugar
$1/3$ C flour
$1/4$ t salt
$1/4$ t baking soda
$2/3$ C quick cooking oats
1 t vanilla extract

Sauce:

Mix 2 T instant vanilla pudding powder with 1 C light cream

Pour peach batter in bottom of skillet.
Crumble topping on peaches.
Cover with foil, top with foiled Jackson Grill.
Cook in warm pit (no coals under skillet). Bank coals around sides and place coals on top of foiled Jackson Grill. Cook for 35 minutes or until topping is light brown. Serve with sauce
.

Walnut and Cranberry Apple Crumble

Topping:

3/4 C all purpose flour
1/4 C light brown sugar
1/4 t cinnamon
1/2 C walnut halves and pieces, coarsely chopped
1/2 stick (1/4 cup) very soft butter, cut into 1/2 inch pieces

Variation: Stir 1 t powdered ginger into topping

Filling:

1 pkg. (12 oz.) frozen cranberries
2 lb. apples, partially peeled, cored, cut into 1/2 inch cubes
1 C sugar
2 T apple juice
1 T cornstarch
1/2 C walnut halves and pieces

Stir together first four topping ingredients in a bowl. Crumble in butter with fingertips until large clumps form.

Cook first four filling ingredients in a large skillet over medium heat, stirring often until mixture gets juicy.

Add cornstarch and boil about 5 minutes until thickened slightly. Stir in walnuts.

Sprinkle with topping. Cover with foil and top with foiled Jackson Grill. Cook in warm pit, (no coals under skillet). Bank sides with medium coals and place coals on top of Jackson Grill.

Cook for 35 minutes until filling is bubbly and topping gets brown. Check and rotate skillet after 15 minutes.

Cast Iron Cooking on the Trail

Raspberry Topped Lemon Pie

Filling:

3 egg yolks
1 (14 oz.) Eagle condensed sweetened milk
$^1/_2$ C lemon juice

Topping:

1 pkg. (12 oz.) frozen raspberries, thawed
1T cornstarch

Mix egg yolks, condensed sweetened milk and lemon juice. Pour into (store bought) graham cracker crust.

If cooking with coals, place pie on top of inverted pie plate in Dutch oven. Cover with foil. Put lid on Dutch oven.

Place medium coals around and on top of oven. Cook until pie is firm, about 35 minutes.

Mix room temperature raspberries with 1 T Cornstarch. Heat until thickened. Spoon raspberry mixture on top of pie.

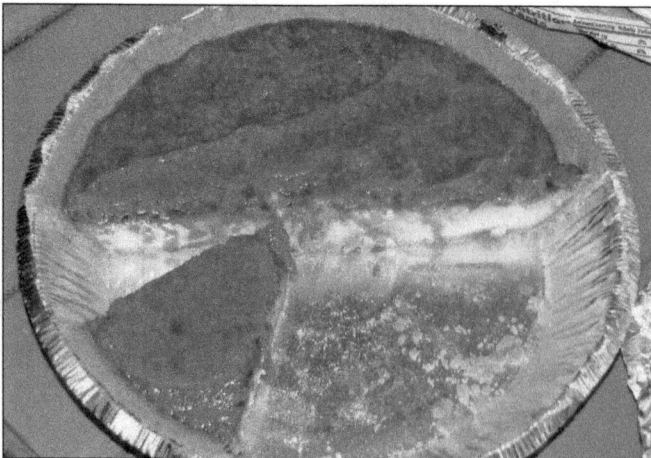

238

Lemon Cake

1 pkg. Lemon Supreme cake mix
4 eggs
1 pkg. lemon Jell-O (3 oz.) gelatin
3/4 C Wesson oil
3/4 C water

Topping:

2 C sifted powdered sugar
$1/2$ C fresh lemon juice or bottled

Mix cake mix, eggs, Jell-O, oil, and water together until smooth, about 2 minutes.

Pour into medium skillet. Cover with foil and place foiled Jackson Grill on top.

Start with warm pit (no coals underneath skillet), bank sides with medium coals and put coals on top.

Bake for 40 minutes or until cake is firm.

Check at 20 and rotate.

Topping: Remove cake from pit and while still hot, use fork to poke the cake full of holes. Stir lemon juice and powdered sugar together until sugar dissolves and pour over hot cake.

Pound Cake
with
Blueberry Orange Sauce

Slice two store bought pound cakes in 1 inch slices.
Option: Fry pound cake in small amount of butter
until toasted brown. Pour sauce over each slice.

Sauce:

1 C orange juice
$1/4$ C honey or sugar
$1\frac{1}{2}$ T flour
1 C blueberries, fresh or frozen (thawed)
2 T butter
Heat until bubbly and thick

Option: serve with whipped cream

*This is such an easy dessert; I usually use it on one of the
long travel days.

Cherry Cobbler

1 C sugar
3/4 C flour
2 t baking powder
pinch of salt
2/3 C milk
1 stick butter
2 pkgs. frozen cherries, (12 oz.) thawed

Sift together sugar, flour, baking powder and salt.

Add milk, stir well.

Melt 1 stick butter in 9 x 9 pan or medium skillet.

Pour batter on top of melted butter. Pour fruit on top of batter.

Cover with foil, top with foiled Jackson Grill.

Start in warm pit; (no coals under skillet or pan), bank sides with coals and put a few on top of the Jackson Grill.

Cook until batter is firm, about 35 minutes. Check and rotate after 15 minutes.

Serve with whipped cream.

Easy Peach Cobbler

1 white cake mix
1 cube butter
1 large (1 lb. 13 oz.) can of peaches

Grease 9 X 13 pan or medium skillet.

Pour entire can of peaches, including syrup into pan. Sprinkle with cake mix. Melt butter and pour over cake mix. Cover with foil and top with foiled Jackson Grill.

Start with warm pit (no coals under pan or skillet). Bank medium coals around sides of skillet and on top of Jackson Grill. After 15 minutes, carefully lift Jackson Grill to check on doneness, adjust heat if necessary and rotate skillet. Cook for 35 to 40 minutes until batter is firm.

Skillet Blueberry Cobbler

Crust:

3 C flour
3/4 C sugar
$1^1/2$ t baking powder
3/4 t salt
3/4 C vegetable shortening, cut into small pieces
3 T butter cut into $^1/2$ pieces

Filling:

2 C sugar
$1^1/2$ C flour
$^1/4$ t salt
$1^1/2$ C sour cream
6 large eggs
7 C fresh blueberries or 4 frozen bags (12 oz.) thawed

Topping:

1 C flour
$^1/2$ C sugar
1 t cinnamon
$^1/4$ t salt
$^1/2$ C butter, cut into pieces
Mix until crumbly.

Mix crust ingredients until crumbly. Gather dough into ball and then flatten into large circle. Press over bottom and up sides of 12 inch (medium) cast iron skillet.

Combine sugar, flour, salt and sour cream. Whip eggs in small bowl, add to batter. Fold blueberries into batter. Spoon filling onto crust. Sprinkle with topping.

Cover with foil and top with foiled Jackson Grill.

Place in pit with medium coals; placing coals around sides and top of Jackson Grill. Cook until top is golden and center is cooked (about $1\frac{1}{2}$ hours)

You will want to check every 20 minutes to adjust heat. Rotate skillet and adjust heat.

Fudge Torte

^{1}/2 C sugar
4 large eggs
2 large egg yolks
1 C semisweet chocolate chips
^{3}/4 C butter
^{1}/3 C seedless raspberry preserves
^{1}/3 C flour
^{1}/8 t salt

Butter medium skillet (9 inch) and set aside.

Whip sugar and eggs, about 5 minutes. Set aside.
Heat butter and chocolate in a saucepan until melted.
Remove and stir in preserves. Fold into egg mixture, and then mix in flour and salt.
Pour into skillet, cover with foil and foiled Jackson Grill.
If pit is warm, (no coals under skillet) add medium coals around sides and on top of grill.
Bake for 30 minutes or until firm.
Check after 15 minutes and rotate skillet
Serve chilled with additional raspberries

Cherry Chocolate Delight

2 chocolate cakes mixes
4 eggs
2 2/3 C applesauce
1 large can (1 lb. 13 oz.) cherry pie filling
1 (6 oz.) pkg. Jell-O instant vanilla pudding
3 C milk
1 tub (8 oz.) Cool Whip topping

Spread pie filling evenly on bottom of large skillet.

Mix cake mix with eggs and applesauce and pour over fillings.

Cover skillet with foil and place foiled Jackson Grill on top. Place hot coals on top of Jackson Grill and around skillet.

Check often to adjust heat, rotate skillet. Bake for 45 minutes or until cake is firm.

While still hot, invert cake on platter. Mix pudding according to box directions. Spoon onto cherries in pinwheel design.

Add whipped topping.

POTS, PANS, PONIES & PINES

Triple Layer Mud Pie

3 squares semi-sweet baking chocolate, melted
$1/4$ C canned sweeten condensed milk
1 Oreo (crushed cookie) pie crust
or chocolate crumb (store bought) pie crust
$1/2$ C chopped pecans, toasted
2 C milk (can be instant)
2 pkg (4 oz.) Jell-O chocolate flavor instant pudding
1 tub (8 oz.) Cool Whip whipped topping, thawed and divided

Mix chocolate and condensed milk until well blended.
Pour into crust; sprinkle with pecans.
Pour milk into large bowl. Add dry pudding mixes.
Beat with whisk 2 min. or until well blended.
Spoon $1^1/2$ C of the pudding over pecans in crust.
Add half of the whipped topping to remaining pudding; stir with whisk until well blended.
Spread over pudding layer in crust; top with remaining whipped topping.
Chill 3 hours.
Anything I have to chill I put on the inside of a cooler pannier or shallow part of stream that I dam up with a few rocks.

Cast Iron Cooking on the Trail

Trail Bars

1 C sugar
1 C peanut butter
1 C instant rolled oats
1 C honey
1 C wheat germ
5 C dry cereal (high protein rice, wheat or oats)

Boil the sugar and honey together for one minute.
Add the peanut butter and stir in quickly.
Pour over the dry ingredients into one 9 X 13 inch buttered pan.
Cut into squares after mixture has set up

Pumpkin Bars

2 C sugar
4 eggs
$^1/_4$ C oil
1 can (1 lb. 13 oz.) pumpkin puree, (not pie filling)
2 t cinnamon
2 C bisquick

Cream cheese frosting:

1 (3 oz.) pkg. cream cheese
1 T milk
1 t vanilla
2 C powdered sugar
$^1/_2$ C butter
Combine all ingredients and mix until smooth.

Pour batter into medium skillet or baking pan.
Cover with foil and top with foiled Jackson Grill.
Place warm pit (no coals under skillet or pan).
Bank hot coals around sides.
Place additional coals on top of Jackson Grill.
Bake until firm, about 50 minutes, but check after 25.
Rotate skillet and adjust heat if necessary.
Let cool. Frost with cream cheese frosting.
Cut into bars or small squares.

Cast Iron Cooking on the Trail

Double Layer Pumpkin Pie

4 oz. ($^1/_2$ of 8 oz. pkg) cream cheese, softened
1 C plus 1 T milk, divided
1 T sugar
1 tub (8 oz.) Cool Whip topping, thawed, divided
1 graham cracker pie crust, (store bought)
1 can (15 oz.) pumpkin puree, not pie filling
2 pkg. (4 oz. serving) vanilla flavor instant pudding and pie filling
1 t ground cinnamon
$^1/_2$ t ground ginger
$^1/_4$ t ground cloves

Mix cream cheese, 1 T milk and the sugar in large bowl until well blended. Gently stir in half of the whipped topping. Spread onto bottom of crust.

Pour 1 C milk into large bowl. Add pumpkin, dry pudding mixes and spices.

Beat with whisk 2 minutes or until well blended. Spread over cream cheese layer.

Chill for 2 hours or until set. Serve with dollop of remaining whipped topping.

Oatmeal Cake

Combine:

$1^1/_2$ C boiling water
1 C quick cooking oatmeal
1 C chopped raisins
Set aside

Cream:

$^1/_2$ C shortening or oil
1 C brown sugar
$^1/_2$ C sugar
2 well beaten eggs

Sift together:

$1^1/_2$ C whole wheat flour
1 t cinnamon
1 t baking soda

Mix water, oatmeal and raisins and set aside to cool
Cream shortening, brown and white sugar and eggs.
Sift together flour, cinnamon and soda.
Combine all of above and bake in 13 X 9 inch greased pan or medium skillet.
Cover with foil and put foiled Jackson Grill on top.
Bank in warm pit (no coals beneath pan or skillet) Bake with medium hot coals around the sides and place some on top of the

Jackson Grill. Bake until firm, about 50 minutes, but check after 25.
Rotate pan or skillet.

Frosting:

1 stick butter
1 C brown sugar
$2^1/2$ T canned milk or cream
$^1/2$ C shredded coconut
$^1/2$ C chopped walnuts

Bring butter, sugar and milk to a boil.
Add coconut and nuts and spread on hot cake

Pineapple Upside-down Cake

1 can ($8^1/2$ oz.) sliced pineapple rings
3 T butter
$1/2$ C brown sugar
maraschino cherries
1 yellow cake mix

Drain pineapple, reserving syrup. Halve pineapple slices.

Melt butter in medium skillet. Add brown sugar and 1 T of the reserved pineapple syrup.

Add water to remaining syrup to make $1/2$ cup.

Arrange pineapple in bottom of skillet placing a cherry in center of each ring.

Prepare yellow cake mix according to directions and spread batter on top.

Cover skillet with foil and place foiled Jackson Grill on top.

Start with warm pit (no coals under skillet). Bank medium to hot coals around sides and place additional coals on top of Jackson Grill.

Bake until cake is firm, about 50 minutes.

Check after 20 and rotate skillet.

Cool ten minutes and invert on plate.

Cranberry Pear Cake with Caramel Sauce

2 C flour
2 C sugar
2 T baking powder
2 eggs

1 large can (1 lb. 13oz.) pear halves (with juice)
3 C frozen cranberries, thawed

Caramel sauce:

$^1/_2$ C butter
$^1/_2$ C cream or half and half
$^1/_2$ C white sugar
$^1/_2$ C brown sugar
$1^1/_2$ t vanilla.

Mix flour, sugar, baking powder and eggs.
Fold in cranberries.

Pour into buttered medium skillet. The pears and cranberries sometimes get clumped together so turn pears face down and distribute the cranberries and pears evenly in the skillet.

Cover skillet with foil and put foiled Jackson Grill on top. Bake in warm pit (do not put coals under skillet). Bank medium to hot coals around side of skillet and on top of Jackson Grill.

Bake for 45 to 50 minutes or until firm. Check after 25 minutes and rotate. Serve with caramel sauce

Bring caramel sauce ingredients (except vanilla) to a boil, stirring continually
Remove from heat, add $1^1/_2$ t vanilla.
Serve hot over cake.

Cherry Fudge Goodies

1 box brownie mix
1¹/₂ C shredded coconut
1¹/₂ C chopped candied cherries

Follow brownie mix recipe.

Add coconut and cherries and blend.

Pour batter into greased Dutch oven or small skillet. If using skillet, cover with foiled Jackson Grill.

Start with warm pit, (no coals under skillet or Dutch oven) Bank medium coals around sides and top of Jackson Grill.

Bake until toothpick comes out clean, about 30 minutes. Check at 20 minutes and rotate skillet. Cut into squares and serve.

Cast Iron Cooking on the Trail

POTS, PANS, PONIES & PINES

Breads

Cast Iron Cooking on the Trail

POTS, PANS, PONIES & PINES

It was the summer of 1988 and the Yellowstone Fires. We had put together a friends' trip. The friends' trips were our group of friends who combined their horses, equipment and different mountain talents to make a memorable week in the mountains. Our plan was to ride down the South Fork trail and stay in one camp for four days at Borner Meadow. There were a few fires in the park, so all camps were under a "fire ban." This meant we could not start a wood fire for any reason. When we checked through the park gate, we joked with one of the rangers that we didn't need our own fire; we would just use one of theirs. Park personnel were still in a good mood as the fire had not really blown up yet. That would soon change.

Usually a friends' trip would start at a party in town with everyone throwing out ideas of where to go and who to invite. Joe Tilden and his wife Ally owned Castle Rock Ranch and Castle Rock Outfitting. Joe would pack and wrangle and Ally would cook. We were friends with two other couples that were ranch owners on the upper South Fork. They provided extra crew and gear. I helped with the cooking. My husband was in charge of the solar shower. He joked he would charge fifty cents to soap the women's backs but would do fronts for free. We had a lot of fun. Everyone pitched in so no one really felt they had a huge workload.

On dude horse pack trips, the wranglers lead everyone down the trail with the pack strings. Usually the guests follow the pack string

259

Cast Iron Cooking on the Trail

with the cook and an additional guest wrangler. On our friends' trips we would leave at different times from the trail head so we would ride into camp at different times.

On that particular trip half of us rode in early to set up the camp. Joe rode in with most of us and Ally rode in with friends that had a later start. Joe had been an outfitter for many years so when he got to camp he was a whirlwind of activity. Joe was at the center directing all of us to put up tents, take care of horses, haul water and get the cook tent organized before the sun went down.

My husband and Joe were assembling the propane oven and burners. The fuel line ran outside the tent to the propane bottle. Hank was outside turning the gas on and Joe was inside lighting the burners. It was not a safe situation as Joe had no idea how much gas Hank had turned on. The rest of us were outside the tent covering the saddles and watering the horses. All of a sudden we heard a huge explosion from the cook tent. We ran in that direction fearing the worst. Hank seemed to be okay, but we couldn't see Joe. Joe walked out; we took one look at him and burst into laughter. Joe had blown off his mustache! None of us had ever seen Joe without a moustache and honestly he looked like a ten year old boy. We were relieved that he wasn't seriously injured, but it sure made for a funny story. It has since been retold around many other mountain campfires.

Ally was riding in later that day. We thought we should even have more fun with Joe. We made Joe hide in the tent. Ally was suspicious as he always greeted her when she rode in, helped her with her gear and tied up her horse. We had another good laugh when she saw her husband, minus a moustache, walk out of the tent.

Each trip seems to develop its own story line. This was also the camp of the "Wild Turkey.". We consumed several bottles of the strong liquor, which we affectionately called "the Bird." Many toasts were made and many stories were told. Some of our friends that joined us on that trip are no longer living. Our memories of our mountain adventures have become even more precious as the years pass.

Each night while we were in that camp, we watched the Yellowstone fires burn along the ridge. We rode out under smokey skies. That summer the fires would burn thousands of acres of back country. I rode seventy five miles through Yellowstone Park in 2017.

It is amazing to see the new growth, actually fairly large trees that have come back from the fires.

Ally was the master organizer for our friends' trips. And she still is an excellent cook. I have included Ally's unbeatable bread in this section.

Biscuits, scones and coffee cakes are good for summer trips. Any yeast bread is easier to prepare in hunting camp. Remember to take an oven thermometer in your kitchen gear. You will definitely have a cook stove in hunting camp and you might be lucky enough to have a pannier oven if you go with the right outfitter in the summer.

Me and Joe fishing in Hidden Basin..

Cast Iron Cooking on the Trail

Mom's Biscuits

4 C flour
2 T baking powder to each C of flour
Add 1 heaping T shortening (Crisco) to each C of flour
(milk amount will vary)
pinch of salt

Mix baking powder and cups of flour, add a pinch of salt. Add shortening (Crisco) to the flour flour. Slowly add the milk; just enough to make the dough moist.

Don't over mix; just enough so you can handle the dough without it sticking to your fingers. Add a little more flour if necessary.

Shape into a ball and roll out on a floured surface. Use the top of a glass or tin can to cut out the biscuits.

Place biscuits on bottom of a medium skillet and cover with foil and foiled Jackson Grill.

Put in warm pit (no coals underneath skillet) but bank coals on the sides and on the top of the Jackson Grill. Bake 10 to 12 minutes. Check after 7 and rotate skillet.

Cast Iron Cooking on the Trail

Buttermilk Biscuits

4 C self-rising flour
1 stick butter, cut into $^1/2$ inch cubes
2 C buttermilk
can use powdered buttermilk
(4 T powdered plus 1 C water =1 C buttermilk)

Combine flour and butter in large bowl. Using back of fork, cut butter into flour until mixture resembles coarse meal.

Gradually mix in $1^3/4$ C buttermilk until dough comes together in large clumps.

Gather dough into ball. Pat out on floured surface to $^1/2$ inch thickness. Using a floured $2^1/2$ inch round cutter, cut out biscuits.

Repeat until all dough is used. Bake in ungreased skillet. Cover with foil and foiled Jackson Grill.

If using a fire pit, use hot coals, but do not place any coals under the skillet. Bank the coals around the sides and add to the top of the Jackson Grill.

Bake for 14 minutes, but check after five minutes. Rotate skillet and adjust heat if necessary.

POTS, PANS, PONIES & PINES

Grandma's Scones

2 C Flour
$2^1/2$ T Sugar
$^1/2$ stick butter
1 t baking soda
3/4 t cream of tarter
Fresh buttermilk or buttermilk powder blend

Mix all ingredients until butter is worked into flour. Add enough buttermilk to hold the dough together and a good consistency to roll out.

Divide dough in half, roll out, and then cut into four pie wedges. Fry on a hot griddle. Serve with honey or jam. Can also be sliced in half creating a top and bottom. Serve with butter and syrup.

Variations:

Cheddar-thyme scones:
After combining butter and dry ingredients, stir $1^1/2$ C grated cheddar cheese and 1 T chopped fresh thyme (or 1 t dry thyme) into flour mixture before adding the buttermilk. Sprinkle tops of scones with an additional $^1/2$ C grated Cheddar cheese before baking.

Raisin scones:
Add $^1/4$ C sugar to dry ingredients. After combining butter and flour mixture, stir in 1 C raisins.

Cast Iron Cooking on the Trail

Theo's Make Ahead Coffee Cake

1$^1/_4$ C flour
$^1/_4$ C sugar
$^1/_4$ C shortening
$^2/_3$ C milk
1 T baking powder
1 egg

Topping:

4 T flour
3 T butter
6 T sugar
$^1/_2$ t cinnamon

Mix dry ingredients. Cut in shortening. Add milk and egg. Grease small skillet. Pour in the batter. Sprinkle topping over batter. Cover with plastic. Keep it cool overnight by placing it on top of a cooler pannier. In the morning, remove plastic wrap, cover with foil and foiled Jackson Grill.

Place in warm pit, place a few coals underneath the skillet, bank the sides and add coals on top of the Jackson Grill. Bake for 30 minutes, but check at 15. Rotate skillet and adjust fire if necessary. Coffee cake should be firm.

POTS, PANS, PONIES & PINES

Raisin Coffee Cake

$^1/_2$ C butter
1 C sugar
2 eggs
1 t vanilla
1 C sour cream
2 C all purpose flour
$1^1/_2$ t baking powder
1 t soda
$^1/_4$ t salt

Topping:

1 C chopped walnuts
$^1/_2$ C brown sugar
1 t ground cinnamon
$1^1/_2$ C raisins

Cream together butter and sugar.
Add eggs and vanilla; beat well.
Blend in sour cream. Set aside.
Mix together flour, baking powder, soda and salt.
Stir into creamed mixture and mix well.
Spread half the batter in greased medium skillet.
Put half the topping on top of batter.
Spoon on remaining batter. Sprinkle on remaining topping.
Cover skillet with foil and foiled Jackson Grill.
Place skillet in warm pit, put a few coals underneath.
Bank the sides and add coals on top of the Jackson Grill.
Bake for 40 minutes, but check at 15.
Rotate skillet and adjust heat.

Ally's Unbeatable Bread
*great for hunting camp

2 C water
7 T margarine
2 t salt
1 T sugar
5 C flour
7 T butter

Mix and set aside to dissolve:
$1/4$ C warm water
1 pkg. yeast

Boil water and add margarine.
When the water is lukewarm, add the yeast mixture, salt and sugar.
Then add flour and mix.
Knead until smooth.
Cover with light towel and let rise until it doubles.
Punch down and place in greased loaf pan.
Bake at 350 for about 30 minutes.

Bubble Bread

Medium size Dutch oven (16 inch)

1 bag frozen dinner rolls (approx. 36)
1 T garlic salt
1 T dried oregano
1 T parsley
$1/2$ C melted butter
$1/2$ C Parmesan cheese

Let rolls thaw and cut in half. Heat empty Dutch oven in fire prior to use then grease the inside of Dutch oven.

Mix seasonings and Parmesan cheese together.

Melt butter, coat dough balls in butter and place one layer in Dutch oven.

Sprinkle seasoning mixture over dough.

Layer dough and seasonings as you go.

Top with extra Parmesan cheese.

Place foil over top of oven and put on lid. Let rise until doubled.

Place Dutch oven in warm pit put a few coals underneath, bank the sides and add some on top.

Bake for 45 minutes, but check at 15. Rotate oven.

Cast Iron Cooking on the Trail

Pull Apart Dinner Rolls

1 C milk
2 T sugar
2 T shortening
1 t salt
1 pkg. active dry yeast
^{1}/4 C lukewarm water
1 egg, beaten
3^1/2 to 3^3/4 C all purpose flour

In small saucepan, combine milk, sugar, shortening and salt; heat to lukewarm.

In a large bowl dissolve yeast in warm water.

Add egg and milk mixture.

Gradually stir in enough flour to make a soft dough.

Turn out onto lightly floured surface and knead gently 2 to 3 minutes to make a smooth ball. Knead in just enough remaining flour so dough is no longer sticky.

Place in greased bowl, turning once to grease surface.

Cover to rise in warm place until double in size (about 1 hour).

Punch dough down and turn out on lightly floured surface, allowing it to rest 10 minutes.

To shape rolls:

Roll or pat dough to a 10 X 8 inch rectangle about 3/4 inch thick.

Cut in 2^1/2 X 1 inch strips.

Lightly roll each strip and place in a greased skillet or 15 X 10 X 1 baking pan, leaving about ^{1}/2 inch between each roll.

Cover and let rise until nearly double in size.

Cook in hot oven or coals (400 degrees for 12-15 minutes).

Frozen dough option:

Thaw a 1 lb. loaf of frozen bread dough. Roll or pat into 10 X 5 inch rectangle.

Cut into $2^1/2$ X 1 inch strips.

Lightly roll each strip and place almost touching in skillet or 9 X 9 X 2 inch pan.

Cover and let rise until nearly double in size.

Cook in hot oven or coals (400 degrees for 30 minutes)

Cast Iron Cooking on the Trail

Raisin Bread

$^1/_2$ C shortening
2 eggs
1 scant C sugar
$1^1/_2$ C cooked raisins
1 C juice off raisins
$2^1/_2$ C flour
1 t soda, dissolved in raisin juice
$^1/_2$ t salt
2 t cinnamon
1 t nutmeg
1 t ground cloves

Cover raisins with water. Simmer for a few minutes.

Cream shortening and sugar together.

Add eggs, drain raisins, reserving 1 C juice, flour and spices.

Grease medium skillet, pour in batter cover with foil and foiled Jackson Grill.

Place in warm pit, put a few coals underneath skillet, bank the sides with coals and add a few on top.

Bake for 50 minutes until firm. Check after 25, rotate skillet and adjust heat.

POTS, PANS, PONIES & PINES

Buttery Cornbread

1 $1/3$ C coarse yellow cornmeal
1 C flour
$1/4$ C sugar
2 t baking powder
$3/4$ t salt
1 C plus 2 T buttermilk or buttermilk powder blend
9 T melted butter
1 large egg plus 1 large egg yolk

Mix cornmeal, flour, sugar, baking powder, and salt in large bowl.
Add buttermilk, melted butter and beaten eggs.
Stir until well blended.
Let mixture stand for 30 minutes to absorb liquid.
Pour batter into foil lined medium skillet.
Cover with foil and foiled Jackson Grill.
Place skillet in warm pit, putting a few coals underneath skillet, bank coals around side and put a few on top of Jackson Grill.
Bake bread until browned around edges (about 40 minutes). Check often and rotate skillet.

Variation:

Fry several slices bacon in skillet and remove.
Leave small amount of grease and pour batter on top.
Fold in crumbled bacon and chives.
Bake as usual.

Cast Iron Cooking on the Trail

Mexican Cornbread

1 can (15 oz.) cream corn
1 C cornmeal
$^1/_2$ C oil
2 beaten eggs
$1^1/_2$ sharp cheddar cheese (grated)
$^3/_4$ C milk
$^1/_2$ t soda
1 t salt
1 can (4 oz.) chopped green chilies

Mix everything but cheese and chilies.

Pour half the batter, then $^1/_2$ the chilies and cheese into a foil lined medium skillet.

Repeat with remaining batter, chilies and cheese.

Cover with foil and foiled Jackson Grill. Place skillet in warm pit, put a few coals underneath skillet. Bank coals around sides and add a few to the top of the Jackson Grill.

Bake for 40 or until edges are brown and bread is firm.

Check at 20 minutes and rotate skillet.

Blueberry Breakfast Bread

2 C Bisquick baking mix
$2/3$ C milk
4 T sugar
1 bag (8 oz.) frozen blueberries, thawed

Mix milk and Bisquick mix together thoroughly with fork.
Pat out half the dough to fit the bottom of greased Dutch oven.
Sprinkle 2 T sugar over dough.
Drain blueberries, saving liquid.
Pour blueberries into the Dutch oven and cover with remaining half of dough, which has been patted into $1/2$ thick inch sheet. Place on blueberries.
Sprinkle remaining sugar on top of dough. After tucking in the sides, pour the liquid from the blueberries on top.
Cover with foil and Dutch oven lid.
Place Dutch oven in medium pit, place a few coals underneath the oven, bank around the sides and add a few on top of lid.
Bake for 40 minutes but check after 20. Bread should be light brown and firm to touch.

7-29

Today we had Pancakes it was very very Good. My Dad went fishing my mom watchd him. Sandy + Marty Did Dishs After Dishes Marty gave Kim + I Bare Back lessons. lessons. We road Bare Back on Waron. Waron is the horse I ride. We made lunch and went on a long horse ride. We got to Jump over logs & creeks. When we got back my Dad and I went fishing. Then we had Dinner Sat around the camp fire and told jokes recited potry, sang, rote in here. Jim can relly sing. He has a big deep voice.

Fan Mail!

INDEX

A

Appetizers

B

Breads

Breakfasts

D

Desserts

S

Salads

V

Vegetables & Side Dishes

POTS, PANS, PONIES & PINES